THE MAGNIFICENT CITY HALL SHATTERED BY EARTHQUAKE.
The violence of the earthquake shock stripped the stone and marble sheathing from the modern building of steel construction and left the skeleton standing. This $7,000,000 structure had only recently been completed.

THE
1906
SAN FRANCISCO
EARTHQUAKE AND FIRE

As Told by Eyewitnesses

Edited by
Charles Morris

DOVER PUBLICATIONS, INC.
Mineola, New York

Bibliographical Note

This Dover edition, first published in 2016, is an unabridged republication of Chapters I through XIV of the work originally published by W. E. Scull, in 1906 under the title *The San Francisco Calamity by Earthquake and Fire.* All of the additional halftone plates that depict San Francisco, originally scattered throughout the remainder of the book, have been included at the end of this book.

Library of Congress Cataloging-in-Publication Data

The 1906 San Francisco earthquake and fire : as told by eyewitnesses / edited by Charles Morris.

    pages cm

    "This Dover edition, first published in 2016, is an unabridged republication of Chapters I through XIV of the work originally published by W. E. Scull, in 1906 under the title The San Francisco Calamity by Earthquake and Fire. All of the additional halftone plates that depict San Francisco have been included at the end of the book."—Data view.

    ISBN-13: 978-0-486-80275-6 (paperback)

    ISBN-10: 0-486-80275-2 (paperback)

    1. San Francisco Earthquake and Fire, Calif., 1906. 2. Earthquakes—California—San Francisco—History—20th century. 3. Fires—California—San Francisco—History—20th century. 4. San Francisco (Calif.)—History—20th century. I. Morris, Charles, 1833–1922, editor.

F869.S357S364 2016

979.4'61051—dc23

2015026000

Manufactured in the United States by RR Donnelley
80275201   2016
www.doverpublications.com

Dedicated

To all who suffered from fire and earthquake in California, and to those who nobly aided in relieving their distress.

Copyright, 1906, by W. E. Scull.
PALL OF DEATH OVER THE DOOMED CITY.
From Oakland, across the bay, the wretched thousands who escaped the fiery charnel-house helplessly watched their homes and all their possessions ascend in flames to the dense smoke which filled the sky.

CARRYING BODIES OUT OF THE RUINS.

Soldiers and policemen forced all idlers at the pistol point to aid the work of taking the dead away for burial in order to prevent an epidemic of disease.

PREFACE

EARTHQUAKE and famine, fire and sudden death—these are the destroyers that men fear when they come singly; but upon the unhappy people of California they came together, a hideous quartette, to slay human beings, to blot from existence the wealth that represented prolonged and strenuous effort, to bring hunger and speechless misery to three hundred thousand homeless and terror-stricken people.

The full measure of the catastrophe can probably never be taken. The summary cannot be made amid the panic, the confusion, the removal of ancient landmarks, the complete subversion of the ordinary machinery of society. When chaos comes, as it did in San Francisco, and all the channels of familiar life are closed, and human anguish grows to be intolerable, compilation of statistics is impossible, even if it were not repugnant to the feelings. And when order is once more restored, after the lapse of many weeks, months and perhaps years, the details of the calamity have merged into one undecipherable mass of misery which defies the analyst and the historian. It is the purpose of this book faithfully to record the story of these awful days when years were lived in a moment and to preserve an accurate chronicle of them, not only for the people whose hearts yearn in sympathy to-day, but for their posterity.

Other frightful catastrophes the world has known. The earthquake which dropped Lisbon into the sea in 1755, and in a moment swallowed up twenty-five thousand people, was perhaps more awful than the convulsion which has brought woe to San Francisco. When Krakatoa Mountain, in the Straits of Sunda, in 1883, split asunder and poured across the land a mighty wave, in which thirty-six thousand human beings perished, the results also were more terrible.

The whirlwind of fire which consumed St. Pierre, in the Island of Martinique, and the devastation wrought by Vesuvius a few days previous to that at San Francisco, need not be used for comparison with the latter tragedy, but they may be referred to, that we may recall the fact that this land of ours is not the only one which has suffered.

But since the western hemisphere was discovered there has been in this quarter of the globe no violence of natural forces at all comparable in destructive fury with that which was manifested upon the Pacific coast. The only other calamity at all equalling it, or surpassing it, was the Civil War, and that was the work of the evil passions of man inciting him to slay his brother, while Nature would have had him live in peace.

The earthquake in San Francisco, which crumbled strong buildings as if they were made of paper, would have been terrible enough; but afterward came the horror of fire and of imprisoned men and women burned alive, and now to it was added the suffering of multitudes from hunger and exposure.

Public attention is fixed on the great city; but smaller cities had their days and nights of destruction, horror and misery. Some were almost destroyed. Others were partly ruined, and beyond their borders, over a wide area, the trembling of the earth toppled houses, annihilated property and transformed riches into poverty. The cost in life can be reckoned. The money loss will never be computed, for the appraised value of the wrecked property conveys no notion of the consequences of the almost complete paralysis, for a time, of the commercial operations by means of which men and women earn their bread.

When the weakness and the folly and the sin of men bring woe upon other men, there are plenty of texts for the preacher and no scarcity of earnest preachers. But here is a vast and awful catastrophe that befell from an act of Nature apparently no more extraordinary than the shrinkage of hot metal in the process of cooling. The consequences are terrifying in this case because they involve the habitations of half a million people; but, no doubt, the process goes on somewhere within the earth almost continuously, and it no more involves the theory of malignant Nature than that of an angry God.

If we contemplate it, possibly we may be helped to a profitable estimate of our own relative insignificance. We think, with some notion of our importance, of the thousand million men who live upon the earth; but they are a mere handful of animate atoms in comparison with the surface, to say nothing of the solid contents, of the globe itself.

We are fond of boasting in this latter day of man's marvelous success in subduing the forces of Nature; and, while we are in the midst of exultation over our victories, Nature tumbles the rocks about somewhere within the bowels of the earth, and we have to learn the old lesson that our triumphs have not penetrated farther than to the very outermost rim of the realms of Nature.

A few weak, almost helpless, creatures, we millions of men stand upon the deck of a great ship, which goes rolling through space that is itself incomprehensible, and usually we are so busy with our paltry ambitions, our transgressions, our righteous labors, our prides and hopes and entanglements that we forget where we are and what is our destiny. A direct interposition from a Superior Power, even if it be hurtful to the body, might be required to persuade us to stop and consider and take anew our bearings, so that we may comprehend in some larger degree our precise relations to things. The wisest men have been the most ready to recognize the beneficence of the discipline of affliction. If there were no sorrow, we should be likely to find the school of life unprofitable.

For one thing, the school wherein sorrow is a part of the discipline is that in which is developed human sympathy, one of the finest and most ennobling manifestations of the Love which is, in its essence, divine. In human life there is much that is ignoble, and the race has almost contemptible weakness and insignificance in comparison with the physical forces of the universe.

But man is superior to all these forces in his possession of the power of affection; and in almost the lowest and basest of the race this power, if latent and half lost, may be found and evoked by the spectacle of the suffering of a fellow-creature.

The human family looks on with pity while the homeless and hungry and impoverished Californians endure pangs. Wherever the news went, by the swift processes of electricity, there men and women, some of them, perhaps, hardly knowing where California is, were sorry and willing and eager to help. There are quarrels within the family sometimes, when nation wars with nation, and all love seems to have vanished; but the world is, in truth, akin. "God hath made of one blood all the nations of the earth," and the blood "tells" when suffering comes.

THE PUBLISHERS.

TABLE OF CONTENTS

---

xiii

THE
1906
SAN FRANCISCO
EARTHQUAKE AND FIRE

As Told by Eyewitnesses

CHAPTER I.

San Francisco and Its Terrific Earthquake.

ON the splendid Bay of San Francisco, one of the noblest harbors on the whole vast range of the Pacific Ocean, long has stood, like a Queen of the West on its seven hills, the beautiful city of San Francisco, the youngest and in its own way one of the most beautiful and attractive of the large cities of the United States. Born less than sixty years ago, it has grown with the healthy rapidity of a young giant, outvieing many cities of much earlier origin, until it has won rank as the eighth city of the United States, and as the unquestioned metropolis of our far Western States.

It is on this great and rich city that the dark demon of destruction has now descended, as it fell on the next younger of our cities, Chicago, in 1872. It was the rage of the fire-fiend that desolated the metropolis of the lakes. Upon the Queen City of the West the twin terrors of earthquake and conflagration have descended at once, careening through its thronged streets, its marts of trade, and its abodes alike of poverty and wealth, and with the red hand of devastation sweeping one of the noblest centres of human industry and enterprise from the face of the earth. It is this story of almost irremediable ruin which it is our unwelcome duty to chronicle. But before entering upon this sorrowful task some description of the city that has fallen a prey to two of the earth's chief agents of destruction must be given.

San Francisco is built on the end of a peninsula or tongue of land lying between the Pacific Ocean and the broad San Francisco Bay, a noble body of inland water extending southward for about forty miles and with a width varying from six to twelve miles. Northward this splendid body of water is connected with San Pablo Bay, ten miles long, and the latter with Suisun Bay, eight miles long, the whole forming a grand range of navigable waters only surpassed by the great northern inlet of Puget Sound. The Golden Gate, a channel five miles long, connects this great harbor with the sea, the whole giving San Francisco the greatest commercial advantages to be found on the Pacific coast.

THE EARLY DAYS OF SAN FRANCISCO.

The original site of the city was a grant made by the King of Spain of four square leagues of land. Congress afterwards confirmed this grant. It was an uninviting region, with its two lofty hills and its various lower ones, a barren expanse of shifting sand dunes extending from their feet. The population in 1830 was about 200 souls, about equal to that of Chicago at the same date. It was not much larger in 1848, when California fell into American hands and the discovery of gold set in train the famous rush of treasure-seekers to that far land. When 1849 dawned the town contained about 2,000 people. They had increased to 20,000 before the year ended. The place, with its steep and barren hills and its sandy stretches, was not inviting, but its ease of access to the sea and its sheltered harbor were important features, and people settled there, making it a depot of mining supplies and a point of departure for the mines.

The place grew rapidly and has continued to grow. At first a city of flimsy frame buildings, it became early a prey to the flames, fire sweeping through it three times in 1850 and taking toll of the young city to the value of $7,500,000. These conflagrations swept away most of the wooden houses, and business men began to build more substantially of brick, stone and iron. Yet to-day, for climatic reasons, most of the residences continue to be built of wood. But the slow-burning redwood of the California hillsides is used instead of the inflammable pine, the result being that since 1850 the loss by fire in the residence section of the city has been remarkably small. In 1900 the city contained 50,494 frame and only 3,881 stone and brick buildings, though the tendency to use more durable materials was then growing rapidly.

Before describing the terrible calamity which fell upon this beautiful city on that dread morning of April 18, 1906, some account of the character of the place is very desirable, that readers may know what San Francisco was before the rage of earthquake and fire reduced it to what it is to-day.

THE CHARACTER OF THE CITY.

The site of the city of San Francisco is very uneven, embracing a series of hills, of which the highest ones, known as the Twin Peaks, reach to an elevation of 925 feet, and form the crown of an amphitheatre of lower altitudes. Several of the latter are covered with handsome residences, and afford a magnificent view of the surrounding country, with its bordering bay and ocean, and the noble Golden Gate channel, a river-like passage from ocean to bay of five miles in length and one in width. This waterway is very deep except on the bar at its mouth, where the depth of water is thirty feet.

Since its early days the growth of the city has been very rapid. In 1900 it held 342,782 people, and the census estimate made from figures of the city directory in 1904 gave it then a population of 485,000, probably a considerable exaggeration. In it are mingled inhabitants from most of the nations of the earth, and it may claim the unenviable honor of possessing the largest population of Chinese outside of China itself, the colony numbering over 20,000.

Of the pioneer San Francisco few traces remain, the old buildings having nearly all disappeared. Large and costly business houses and splendid residences have taken their place in the central portion of the city, marble, granite, terra-cotta, iron and steel being largely used as building material. The great prevalence of frame buildings in the residence sections is largely due to the popular belief that they are safer in a locality subject to earthquakes, while the frequent occurrence of earth tremors long restrained the inclination to erect lofty buildings. Not until 1890 was a high structure built, and few skyscrapers had invaded the city up to its day of ruin. They will probably be introduced more frequently in the future, recent experience having demonstrated that they are in considerable measure earthquake proof.

The city before the fire contained numerous handsome structures, including the famous old Palace Hotel, built at a cost of $3,000,000 and with accommodations for 1,200 guests; the nearly finished and splendid Fairmount Hotel; the City Hall, with its lofty dome, on which $7,000,000 is said to have been spent, much of it, doubtless, political plunder; a costly United States Mint and Post Office, an Academy of Science, and many churches, colleges, libraries and other public edifices. The city had 220 miles of paved streets, 180 miles of electric and 77 of cable railway, 62 hotels, 16 theatres, 4 large libraries, 5 daily newspapers, etc., together with 28 public parks.

Sitting, like Rome of old, on its seven hills, San Francisco has long been noted for its beautiful site, clasped in, as it is, between the Pacific Ocean and its own splendid bay, on a peninsula of some five miles in width. Where this juts into the bay at its northernmost point rises a great promontory known as Telegraph Hill, from whose height homeless thousands have recently gazed on the smoke rising from their ruined homes. In the early days of golden promise a watchman was stationed on this hill to look out for coming ships entering the Golden Gate from their long voyage around the Horn and signal the welcome news to the town below. From this came its name.

Cliffs rise on either side of the Golden Gate, and on one is perched the Cliff House, long a famous hostelry. This stands so low that in storms the surf is flung over its lower porticos, though its force is broken by the Seal Rocks. A chief attraction to this house was to see the seals play on these rocks, their favorite place of resort. The Cliff House was at first said to have been swept bodily by the earthquake into the sea, but it proved to be very little injured, and stands erect in its old picturesque location.

In the vicinity of Telegraph Hill are Russian and Nob Hills, the latter getting its peculiar title from the fact that the wealthy "nobs," or mining magnates, of bonanza days built their homes on its summit level. Farther to the east are Mount Olympus and Strawberry Hill, and beyond these the Twin Peaks, which really embrace three hills, the third being named Bernal Heights. Farther to the south and east is Rincan Hill, the last in the half moon crescent of hills, within which is a spread of flat ground extending to the bay. Behind the hills on the Pacific side stretches a vast sweep of sand, at some places level, but often gathered into great round dunes. Part of this has been transformed into the beautiful Golden Gate Park, a splendid expanse of green verdure which has long been one of San Francisco's chief attractions.

Beneath the whole of San Francisco is a rock formation, but everywhere on top of this extends the sand, the gift of the winds. This is of such a character that a hole dug in the street anywhere, even if only to the depth of a few feet, must be shored up with planking or it will fill as fast as it is excavated, the sand running as dry as the contents of an hour glass. When there is an earthquake—or a "temblor," to use the Spanish name—it is the rock foundation that is disturbed, not the sand, which, indeed, serves to lessen the effect of the earth tremor.

THE FOUNDATIONS OF THE CITY.

Leaving the region of the hills and descending from their crescent-shaped expanse, we find a broad extent of low ground, sloping gently toward the bay. On this low-lying flat was built all of San Francisco's business houses, all its principal hotels and a large part of its tenements and poorer dwellings. It was here that the earthquake was felt most severely and that the fire started which laid waste the city.

Rarely has a city been built on such doubtful foundations. The greater part of the low ground was a bay in 1849, but it has since been filled in by the drifting sands blown from the ocean side by the prevailing west winds and by earth dumped into it. Much of this land was "made ground." Forty-niners still alive say that when they first saw San Francisco the waters of the bay came up to Montgomery Street. The Palace Hotel was in Montgomery Street, and from there to the ferry docks—a long walk for any man—the water had been driven back by a "filling-in" process.

This is the district that especially suffered, that south of Market and east of Montgomery Streets. Nearly all the large buildings in this section are either built on piles driven into the sand and mud or were raised upon wooden foundations. It is on such ground as this that the costly Post Office building was erected, despite the protests of nearly the entire community, who asserted that the ground was nothing but a filled-in bog.

In none of the earthquakes that San Francisco has had was any serious damage except to houses in this filled-in territory, and to houses built along the line of some of the many streams which ran from the hills down to the bay, and which were filled in as the town grew—for instance, the Grand Opera House was built over the bed of St. Anne's Creek. A bog, slough and marsh, known as the Pipeville Slough, was the ground on which the City Hall was built, and which was originally a burying ground. Sand from the western shore had blown over and drifted into the marsh and hardened its surface.

When the final grading scheme of the city was adopted in 1853, and work went on, the water front of the city was where Clay Street now is, between Montgomery and Sansome Streets. The present level area of San Francisco of about three thousand acres is an average of nine feet above or below the natural surface of the ground and the changes made necessitated the transfer of 21,000,000 cubic yards from hills to hollows. Houses to the number of thousands were raised or lowered, street floors became subcellars or third stories and the whole natural face of the ground was altered.

Through this infirm material all the pipes of the water and sewer system of San Francisco in its business districts and in most of the region south of Market street were laid. When the earthquake came, the filled-in ground shook like the jelly it is. The only firm and rigid material in its millions of cubic yards of surface area and depth were the iron pipes. Naturally they broke, as they would not bend, and San Francisco's water system was therefore instantly disabled, with the result that the fire became complete master of the situation and raged uncontrolled for three days and nights.

Although the earthquake wrecked the business and residential portions of the city alike, on the hills the land did not sink. All "made ground" sank in consequence of the quaking, but on the high ground the upper parts of the buildings were about the only portions of the structures wrecked. Most of the damage on the hills was done by falling chimneys. On Montgomery Street, half a block from the main office of the Western Union Company, the middle of the street was cracked and blown up, but during the shocks which struck the Western Union building only the top stories were cracked. Similar phenomena were experienced in other localities, and the bulk of the disaster, so far as the earthquake was concerned, was confined to the low-lying region above described.

THE BANE OF THE EARTHQUAKE.

From the origin of San Francisco the earthquake has been its bane. During the past fifty years fully 250 shocks have been re-

THOUSANDS FIGHTING THEIR WAY TO THE FERRY.

Refugees from the crowded portions of the stricken city dragged their belongings through the wreckage in the streets toward the ferry house, only to find that boats were lacking to carry them across to Oakland.

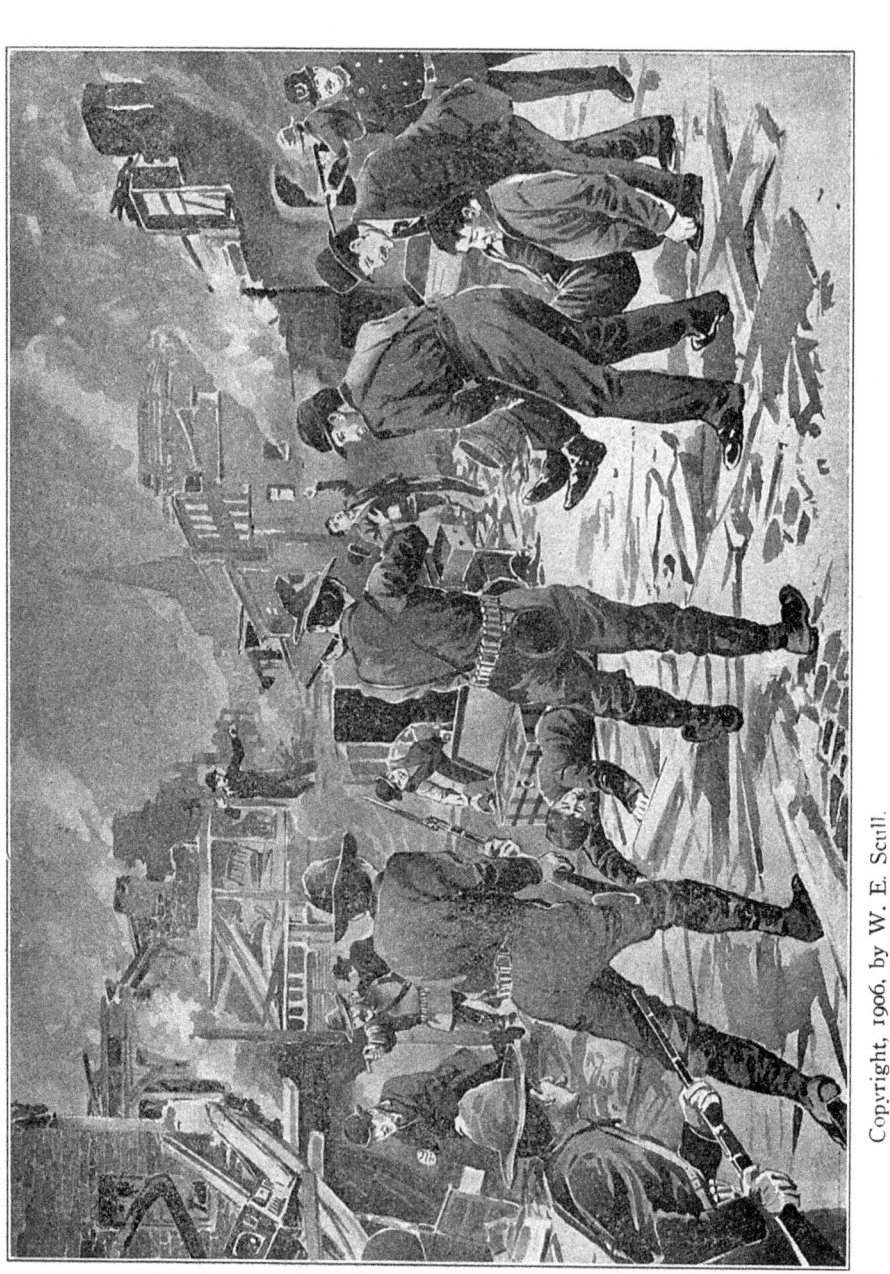

Copyright, 1906, by W. E. Scull.
LOOTERS AND ROBBERS SHOT BY THE MILITARY.
Mayor Schmitz issued orders to the soldiers that all malefactors and robbers were to be shot on sight. During the fire an attempt was made to rob the Mint. and the thieves were killed.

corded, while all California has been subject to them.   But fre-
quency rather than violence of shocks has been the characteristic
of the seismic history of the State, there having been few shocks
that caused serious damage, and none since 1872 that led to loss
of life.

There was a violent shock in 1856, when the city was only a
mining town of small frame buildings.   Several shanties were overthrown and a few persons killed by falling walls and chimneys.
There was a severe shock also in 1865, in which many buildings
were   shattered.   Next in violence was the shock of 1872, which
cracked the walls of some of the public buildings and caused a panic.
There was no great loss of life.   In April, 1898, just before midnight, there was a lively shakeup which caused the tall buildings to
shake like the snapping of a whip and drove the tourists out of the
hotels into the streets in their nightclothes.   Three or four old
houses fell, and the Benicia Navy Yard, which is on made ground
across the bay, was damaged to the extent of about $100,000.   The
last severe shock was in January, 1900, when the St. Nicholas Hotel
was badly damaged.

These were the heaviest shocks.   On the other hand, light
shocks, as above said, have been frequent.   Probably the sensible
quakes have averaged three or four a year.   These are usually
tremblings lasting from ten seconds to a minute and just heavy
enough to wake light sleepers or to shake dishes about on the shelves.
Tourists and newcomers are generally alarmed by these phenomena,
but old Californians have learned to take them philosophically.   To
one who is not afraid of them, the sensation of one of these little
tremblers is rather pleasant than otherwise, and the inhabitants

grew so accustomed to them as rarely to let them disturb their equanimity.

After 1900 the forces beneath the earth seemed to fall asleep. As it proved, they were only biding their time. The era was at hand when they were to declare themselves in all their mighty power and fall upon the devoted city with ruin in their grasp. But all this lay hidden in the secret casket of time, and the city kept up to its record as one of the liveliest and in many respects the most reckless and pleasure-loving on the continent, its people squandering their money with thoughtless improvidence and enjoying to the full all the good that life held out to them.

On the 17th of April, 1906, the city was, as usual, gay, careless, busy, its people attending to business or pleasure with their ordinary vim as inclination led them, and not a soul dreaming of the horrors that lay in wait. They were as heedless of coming peril and death as the inhabitants of Sodom and Gomorrah before the rain of fire from heaven descended upon their devoted heads. This is not to say that they were doomed by God to destruction like these "cities of the plains." We should more wisely say that the forces of ruin within the earth take no heed of persons or places. They come and go as the conditions of nature demand, and if man has built one of his cities across their destined track, its doom comes from its situation, not from the moral state of its inhabitants.

THE GREAT DISASTER OF 1906.

That night the people went, with their wonted equanimity, to their beds, rich and poor, sick and well alike. Did any of them dream of disaster in the air? It may be so, for often, as the poet tells us, "Coming events cast their shadows before." But, forewarned by dreams or not, doubtless not a soul in the great city was prepared for the terrible event so near at hand, when, at thirteen minutes past five o'clock on the dread morning of the 18th, they felt their beds lifted beneath them as if by a Titan hand, heard the crash of falling walls and ceilings, and saw everything in their rooms tossed madly about, while through their windows came the roar of an awful disaster from the city without.

It was a matter not of minutes, but of seconds, yet on all that coast, long the prey of the earthquake, no shock like it had ever been felt, no such sudden terror awakened, no such terrible loss occasioned as in those few fearful seconds. Again and again the trembling of the earth passed by, three quickly repeated shocks, and the work of the demon of ruin was done. People woke with a start to find themselves flung from their beds to the floor, many of them covered with the fragments of broken ceilings, many lost among the ruins of falling floors and walls, many pinned in agonizing suffering under the ruins of their houses, which had been utterly wrecked in those fatal seconds. Many there were, indeed, who had been flung to quick if not to instant death under their ruined homes.

Those seconds of the reign of the elemental forces had turned the gayest, most careless city on the continent into a wreck which no words can fitly describe. Those able to move stumbled in wild panic across the floors of their heaving houses, regardless of clothing, of treasures, of everything but the mad instinct for safety, and rushed headlong into the streets, to find that the earth itself had yielded to the energy of its frightful interior forces and had in places been torn and rent like the houses themselves. New terrors assailed the fugitives as fresh tremors shook the solid ground, some of them strong enough to bring down shattered walls and chimneys, and bring back much of the mad terror of the first fearful quake. The heaviest of these came at eight o'clock. While less forcible than that which had caused the work of destruction, it added immensely to the panic and dread of the people and put many of the wanderers to flight, some toward the ferry, the great mass in the direction of the sand dunes and Golden Gate Park.

The spectacle of the entire population of a great city thus roused suddenly from slumber by a fierce earthquake shock and sent flying into the streets in utter panic, where not buried under falling walls or tumbling debris, is one that can scarcely be pictured in words, and can be given in any approach to exact realization only in the narratives of those who passed through its horrors and experienced the sensations to which it gave rise. Some of the more vivid of these personal accounts will be presented later, but at present we must confine ourselves to a general statement of the succession of events.

The earthquake proved but the beginning and much the least destructive part of the disaster. In many of the buildings there were fires, banked for the night, but ready to kindle the inflammable material hurled down upon them by the shock. In others were live electric wires which the shock brought in contact with woodwork. The terror-stricken fugitives saw, here and there, in all directions around them, the alarming vision of red flames curling upward and outward, in gleaming contrast to the white light of dawn just showing in the eastern sky. Those lurid gleams climbed upward in devouring haste, and before the sun had fairly risen a dozen or more conflagrations were visible in all sections of the business part of the city, and in places great buildings broke with startling suddenness into flame, which shot hotly high into the air.

CONSPICUOUS FIGURES IN THE RELIEF WORK.

CLAUS SPRECKELS, the Sugar King, who suffered enormous losses and yet contrbuted freely to alleviate suffering.

GENERAL FUNSTON, who prevented a reign of terror by strict military discipline.

Secretary of Commerce and Labor METCALF, who was interested in the work of relief.

CELEBRATED BUILDINGS OF SAN FRANCISCO.
The upper picture is a photograph of the Dolores Mission. Below, the
United States Mint, which was saved from the flames.

While the mass of the people were stunned by the awful suddenness of the disaster and stood rooted to the ground or wandered
helplessly about in blank dismay, there were many alert and self-
possessed among them who roused themselves quickly from their
dismay and put their energies to useful work.   Some of these gave
themselves to the work of rescue, seeking to save the injured from
their perilous situation and draw the bodies of the dead from the
ruins under which they lay.   Those base wretches to whom plunder
is always the first thought were as quickly engaged in seeking for
spoil in edifices laid open to their plundering hands by the shock.
Meanwhile the glare of the flames brought the fire-fighters out in
hot haste with their engines, and up from the military station at the
Presidio, on the Golden Gate side of the city, came at double quick
a force of soldiers, under the efficient command of General Funston,
of Cuban and Philippine fame.   These trained troops were at once
put on guard over the city, with directions to keep the best order
possible, and with strict command to shoot all looters at sight.
Funston recognized at the start the necessity of keeping the lawless
element under control in such an exigency as that which he had to
face.   Later in the day the First Regiment of California National
Guards was called out and put on duty, with similar orders.

RESCUERS AND FIRE-FIGHTERS.

The work of fighting the fire was the first and greatest duty to
be performed, but from the start it proved a very difficult, almost a
hopeless, task.   With fierce fires burning at once in a dozen or more
separate places, the fire department of the city would have been
inadequate to cope with the demon of flame even under the best of
circumstances.   As it was, they found themselves handicapped at

the start by a nearly total lack of water. The earthquake had disarranged and broken the water mains and there was scarcely a drop of water to be had, so that the engines proved next to useless. Water might be drawn from the bay, but the centre of the conflagration was a mile or more away, and this great body of water was rendered useless in the stringent exigency.

The only hope that remained to the authorities was to endeavor to check the progress of the flames by the use of dynamite, blowing up buildings in the line of progress of the conflagration. This was put in practice without loss of time, and soon the thunder-like roar of the explosions began, blasts being heard every few minutes, each signifying that some building had been blown to atoms. But over the gaps thus made the flames leaped, and though the brave fellows worked with a desperation and energy of the most heroic type, it seemed as if all their labors were to be without avail, the terrible fire marching on as steadily as if a colony of ants had sought to stay its devastating progress.

THE HORROR OF THE PEOPLE.

It was with grief and horror that the mass of the people gazed on this steady march of the army of ruin. They were seemingly half dazed by the magnitude of the disaster, strangely passive in the face of the ruin that surrounded them, as if stunned by despair and not yet awakened to a realization of the horrors of the situation. Among these was the possibility of famine. No city at any time carries more than a few days' supply of provisions, and with the wholesale districts and warehouse regions invaded by the flames the shortage of food made itself apparent from the start. Water was even more difficult to obtain, the supply being nearly all cut off. Those who possessed supplies of food and liquids of any kind in many cases took advantage of the opportunity to advance their prices. Thus an Associated Press man was obliged to pay twenty-five cents for a small glass of mineral water, the only kind of drink that at first was to be had, while food went up at the same rate, bakers frequently charging as much as a dollar for a loaf. As for the expressmen and cabmen, their charges were often practically prohibitory, as much as fifty dollars being asked for the conveyance of a passenger to the ferry. Policemen were early stationed at some of the retail shops, regulating the sale and the price of food, and permitting only a small portion to be sold to each purchaser, so as to prevent a few persons from exhausting the supply.

The fire, the swaying and tottering walls, the frequent dynamite explosions, each followed by a crashing shower of stones and bricks, rendered the streets very unsafe for pedestrians, and all day long the flight of residents from the city went on, growing quickly to the dimensions of a panic. The ferryboats were crowded with those who wished to leave the city, and a constant stream of the homeless, carrying such articles as they had rescued from their homes, was kept up all day long, seeking the sand dunes, the parks and every place uninvaded by the flames. Before night Golden Gate Park and the unbuilt districts adjoining on the ocean side presented the appearance of a tented city, shelter of many kinds being improvised from bedding and blankets, and the people settling into such sparse comfort as these inadequate means provided.

A strange feature of the disaster was a rush to the banks by people who wished to get their money and flee from the seemingly doomed city. The fire front was yet distant from these institutions, which were destined to fall a prey to the flames, and all that morning lines of dishevelled and half-frantic men stood before the banks on Montgomery and Sansome Strets, braving in their thirst for money the smoke and falling embers and beating in wild anxiety upon the doors. Their effort was vain; the doors remained closed; finally the police drove these people away, and the banks went on with the work of saving their valuables. As for the people who wildly fled toward the ferries, in spite of the fact that ten blocks of fire, as the day went on, stopped all egress in that direction, it became necessary for them to be driven back by the police and the troops, and they were finally forced to seek safety in the sands. And thus, with incident manifold, went on that fatal Wednesday, the first day of the dread disaster.

OFFICIAL RECORD OF THE EARTHQUAKE.

It is important here to give the official record of the earthquake shocks, as given by the scientists. Professor George Davidson, of the University of California, says of them:

"The earthquake came from north to south, and the only description I am able to give of its effect is that it seemed like a terrier shaking a rat. I was in bed, but was awakened by the first shock. I began to count the seconds as I went towards the table where my watch was, being able through much practice closely to approximate the time in that manner. The shock came at 5.12 o'clock. The first sixty seconds were the most severe. From that time on it decreased gradually for about thirty seconds. There was then the slightest perceptible lull. Then the shock continued for sixty seconds longer, being slighter in degree in this minute than in any part of the preceding minute and a half. There were two slight shocks afterwards which I did not time. At 8.14 o'clock I recorded a shock of five seconds' duration, and one at 4.15 of two seconds. There were slight shocks which I did not record at 5.17 and at 5.27. At 6.50 P. M. there was a sharp shock of several seconds."

Professor A. O. Louschner, of the students' observatory of the University of California, thus records his observations:

"The principal part of the earthquake came in two sections, the first series of vibrations lasting about forty seconds. The vibrations diminished gradually during the following ten seconds, and then occurred with renewed vigor for about twenty-five seconds more. But even at noon the disturbance had not subsided, as slight shocks are recorded at frequent intervals on the seismograph. The motion was from south-southeast to north-northwest.

"The remarkable feature of this earthquake, aside from its intensity, was its rotary motion. As seen from the print, the sum total of all displacements represents a very regular ellipse, and some of the lines representing the earth's motion can be traced along the whole circumference. The result of observation indicates that our heaviest shocks are in the direction south-southeast to north-northwest. In that respect the records of the three heaviest earthquakes agree entirely. But they have several other features in common. One of these is that while the displacements are very large the vibration period is comparatively slow, amounting to about one second in the last two big earthquakes."

If we seek to discover the actual damage done by the earthquake, the fact stands out that the fire followed so close upon it that the traces of its ravages were in many cases obliterated. So many buildings in the territory of the severest shock fell a prey to the flames or to dynamite that the actual work of the earth forces was made difficult and in many places impossible to discover. This fact is likely to lead to considerable dispute and delay when the question of insurance adjustment comes up, many of the insurance companies confining their risk to fire damage and claiming exemption from liability in the case of damage due to earthquake.

Among the chief victims of the earth-shake was the costly and showy City Hall, with its picturesque dome standing loftily above the structure. This dome was left still erect, but only as a skeleton might stand, with its flesh gone and its bare ribs exposed to the searching air. Its roof, its smaller towers came tumbling down in frightful disarray, and the once proud edifice is to-day a miserable wreck, fire having aided earthquake in its ruin. The new Post Office, a handsome government building, also suffered severely from the shock, its walls being badly cracked and injury done by earthquake and fire that it is estimated will need half a million dollars to repair.

FREAKS OF THE EARTHQUAKE.

One observer states that the earthquake appeared to be very irregular in its course. He tells us that "there are gas reservoirs with frames all twisted and big factories thrown to the ground, while a few yards away are miserable shanties with not a board out of place. Wooden, steel and brick structures hardly felt the earthquake in some parts of the city, while in other places all were wrecked.

"Skirting the shore northwest from the big ferry building—which was so seriously injured that it will have to be rebuilt—the first thing observed was the extraordinary irregularity of the earthquake's course. Pier No. 5, for instance, is nothing but a mass of ruins, while Pier No. 3, on one side of it and Pier No. 7, on the other side, similar in size and construction, are undamaged. Farther on, the Kosmos Line pier is a complete wreck."

The big forts at the entrance to the Golden Gate also suffered seriously from the great shake-up, and the emplacements of the big guns were cracked and damaged. The same is the case with the fortifications back of Old Fort Point, the great guns in these being for the present rendered useless. It will take much time and labor to restore their delicate adjustment upon their carriages.

The buildings that collapsed in the city were all flimsy wooden buildings and old brick structures, the steel frame buildings, even the score or more in course of construction, escaping injury from the earthquake shock. Of the former, one of the most complete wrecks was the Valencia Hotel, a four-story wooden building, which collapsed into a heap of ruins, pinning many persons under its splintered timbers.

SKYSCRAPERS EARTHQUAKE PROOF.

In fact, as the reports of damage wrought by the earthquake came in, the conviction grew that one of the safest places during the earthquake shock was on one of the upper floors of the sky-scraper office buildings or hotels. As a matter of fact, not a single person, so far as can be learned, lost his or her life or was seriously injured in any of the tall, steel frame structures in the city, although they rocked during the quake like a ship in a gale.

The loss of life was caused in almost every case by the collapse of frame structures, which the native San Franciscan believed was the safest of all in an earthquake, or by the shaking down of portions of brick or stone buildings which did not possess an iron framework. The manner in which the tall steel structures withstood the shock is a complete vindication of the strongest claims yet made for them, and it is made doubly interesting from the fact that this is the first occasion on which the effect of an earthquake of any proportions on a tall steel structure could be studied.

The St. Francis Hotel, a sixteen-story structure, can be repaired at an expenditure of about $400,000, its damage being almost wholly by fire. The steel shell and the floors are intact. Although the building rocked like a ship in a gale while the quake lasted, its foundations are undamaged. Other steel buildings which are so little damaged as to admit of repairs more or less extensive are the James Flood, the Union Trust, the *Call* building, the Mutual Savings Bank, the Crocker-Woolworth building and the Postal building. All of these are modern buildings of steel construction, from sixteen to twenty stories.

A peculiar feature of the effect of the earthquake on structures of this kind is reported in the case of the Fairmount Hotel, a fourteen-story structure. The first two stories of the Fairmount are found to be so seriously damaged that they will have to be rebuilt, while the other twelve stories are uninjured.

Various explanations are being made of the surprising resistance shown by the skyscrapers. The great strength and binding power of the steel frame, combined with a deep-seated foundation and great lightness as compared with buildings of stone, are the main reasons given. The iron, it is said, unlike stone, responded to the vibratory force and passed it along to be expended in other directions, while brick or stone offered a solid and impenetrable front, with the result that the seismic force tended to expend itself by shaking the building to pieces.

Whether there is any scientific basis for the latter theory or not, it seems reasonable enough, in view of the descriptions given us of the manner in which the steel buildings received the shock. All things considered, the modern steel building has afforded in the San Francisco earthquake the most convincing evidence of its strength.

From Golden Gate Park came news of the total destruction of the large building covering a portion of the children's playground. The walls were shattered beyond repair, the roof fell in, and the destruction was complete. The pillars of the new stone gates at the park entrance were twisted and torn from their foundations, some of them, weighing nearly four tons, being shifted as though they were made of cork. It is a little singular that the monuments and statues in the city escaped without damage except in the case of the imposing Dewey Monument, in Union Square Park, which suffered what appears to be a minor injury.

In this connection an incident of extraordinary character is narrated. Among the statues on the buildings of the Leland Stanford, Jr., University, all of which were overthrown, was a marble statue of Carrara in a niche on the building devoted to zoology and physiology. This in falling broke through a hard cement pavement and buried itself in the ground below, from which it was dug. The singular fact is that when recovered it proved to be without a crack or scratch. This university seemed to be a central point in the disturbance, the destruction of its buildings being almost total, though they had been built with the especial design of resisting earthquake shocks.

Such was the general character of the earthquake at San Francisco and in its vicinity. It may be said farther that all, or very nearly all, the deaths and injuries were due to it directly or indirectly, even those who perished by fire owing their deaths to the fact of their being pinned in buildings ruined by the earthquake shock, while others were killed by falling walls weakened by the same cause.

On the night of April 23d the earth tremor returned with a slight shock, only sufficient to cause a temporary alarm. On the afternoon of the 25th came another and severer one, strong enough to shake down some tottering walls and add another to the list of victims. This was a woman named Annie Whitaker, who was at work in the kitchen of her home at the time. The chimney, which had been weakened by the great shock, now fell, crashing through the roof and fracturing her skull. Thus the earth powers claimed a final human sacrifice before their dread visitation ended.

CHAPTER II.

The Demon of Fire Invades the Stricken City.

THE terrors of the earthquake are momentary. One fierce, levelling shock and usually all is over. The torment within the earth has passed on and the awakened forces of the earth's crust sink into rest again, after having shaken the surface for many leagues. Rarely does the dread agent of ruin leave behind it such a terrible follower to complete its work as was the case in the doomed city of San Francisco. All seemed to lead towards such a carnival of ruin as the earth has rarely seen. The demon of fire followed close upon the heels of the unseen fiend of the earth's hidden caverns, and ran red-handed through the metropolis of the West, kindling a thousand unhurt buildings, while the horror-stricken people stood aghast in terror, as helpless to combat this new enemy as they were to check the ravages of the earthquake itself.

Why not quench the fire at its start with water? Alas! there was no water, and this expedient was a hopeless one. The iron mains which carried the precious fluid under the city streets were broken or injured so that no quenching streams were to be had. In some cases the engine houses had been so damaged that the fire-fighting apparatus could not be taken out, though even if it had it would have been useless. A sweeping conflagration and not an ounce of water to throw upon it! The situation of the people was a maddening one. They were forced helplessly and hopelessly to gaze upon the destruction of their all, and it is no marvel if many of them grew frantic and lost their reason at the sight. Thousands gathered and looked on in blank and pitiful misery, their strong hands, their iron wills of no avail, while the red-lipped fire devoured the hopes of their lives.

In a dozen, a hundred, places the flames shot up redly. Huge, strong buildings which the earthquake had spared fell an unresisting prey to the flames. The great, iron-bound, towering Spreckles building, a steeple-like structure, of eighteen stories in height, the tallest skyscraper in the city, had resisted the earthquake and remained proudly erect. But now the flames gathered round and assailed it. From both sides came their attack. A broad district near by, containing many large hotels and lodging houses, was being fiercely burnt out, and soon the windows of the lofty building cracked and splintered, the flames shot triumphantly within, and almost in an instant the vast interior was a seething furnace, the wild flames rushing and leaping within until only the blackened walls remained.

THE RESISTLESS MARCH OF THE FLAMES.

This was the region of the newspaper offices, and they quickly succumbed. The *Examiner,* standing across Third Street from Spreckles, collapsed from the earthquake shock. A flimsy edifice, it had long been looked upon as dangerous. Another building in the rear of this alone resisted both flames and smoke. Across Market Street from the *Examiner* stood the *Chronicle* building, a dozen stories high. Firmly built, it had borne the earthquake assault unharmed, but the flames were an enemy against which

FIRE RAVAGING MARKET STREET,
The principal street of San Francisco's business centre. It was one of the first
quarters to be devastated by the flames. House after house succumbed,
but the lofty "Call", or Spreckels building remained standing.

ESCAPING THE HAVOC IN TRAINS.

The station at Oakland was thronged by refugees who were given free transportation to different parts of California by the railroads. Some men found places on the roofs of crowded cars.

PHOTOGRAPHED FROM A HOUSE-TOP IN SAN FRANCISCO.
Portraying the terrifying sweep of the flames better than the pen can picture, this illustration shows the progress of the destroying element at work in the theatre district of San Francisco. The scene is just off Market Street and looking up O'Farrell Street. The ruins of the Columbia Theatre loom up prominently.

Copyright, 1906, by O. F. Browning.

FLEEING FROM THE FLAMES IN SAN FRANCISCO.

This photograph shows thousands of the homeless wandering through Broadway. Telegraph Hill is seen in the distance. In the centre of the picture a woman is seen carrying her baby and pushing a baby-carriage full of her belongings. To the right appears a man taking an injured woman to a place of safety.

it had no defense, and it was quickly added to the victims of the fire-fiend.

Farther down Market Street, the chief business thoroughfare of the city, stood that great caravansary, the Palace Hotel, which for thirty years had been a favorite hostelry, housing the bulk of the visitors to the Californian metropolis. Its time had come. Doom hovered over it. Its guests had fled in good season, as they saw the irresistible approach of the conquering flames. Soon it was ablaze; quickly from every window of its broad front the tongues of flame curled hotly in the air; it became a thrice-heated furnace, like so many of the neighboring structures, adding its quota to the vast cloud of smoke that hung over the burning city, and rapidly sinking in red ruin to the earth.

All day Wednesday the fire spread unchecked, all efforts to stay its devouring fury proving futile. In the business section of the city everything was in ruins. Not a business house was left standing. Theatres crumbled into smouldering heaps. Factories and commission houses sank to red ruin before the devouring flames. The scene was like that of ancient Babylon in its fall, or old Rome when set on fire by Nero's command, as tradition tells. In modern times there has been nothing to equal it except the conflagration at Chicago, when the flames swept to ruin that queen city of the Great Lakes.

When night fell and the sun withdrew his beams the spectacle was one at once magnificent and awe-inspiring. The city resembled one vast blazing furnace. Looking over it from a high hill in the western section, the flames could be seen ascending skyward for miles upon miles, while in the midst of the red spirals of flame could be seen at intervals the black skeletons and falling towers of doomed buildings.   Above all this hung a dense pall of smoke, showing lurid where the flames were reflected from its dark and threatening surface.   To those nearer the scene presented many pathetic and distressing features, the fire glare throwing weird shadows over the worn and panic-stricken faces of the woe-begone fugitives, driven from their homes and wandering the streets in helpless misery.   Many of them lay sleeping on piles of blankets and clothing which they had brought with them, or on the hard sidewalks, or the grass of the open parks.

THE CARE OF THE WOUNDED.

Through all the streets ambulances and express wagons were hurrying, carrying dead and injured to morgues and hospitals.   But these refuges for the wounded or receptacles for the dead were no safer than the remainder of the city.   In the morgue at the Hall of Justice fifty bodies lay, but the approach of the flames rendered it necessary to remove to Jackson Square these mutilated remnants of what had once been men.   Hospitals were also abandoned at intervals, doctors and nurses being forced to remove their patients in haste from the approaching flames.

There is an open park opposite City Hall.   Here the Board of Supervisors met, and, with fifty substantial citizens who joined them, formed a Committee of Safety, to take in hand the direction of affairs and to seek safe quarters for the dying and the dead.   Strangely enough, Mechanics' Pavilion, opposite City Hall, had escaped injury from the earthquake, though it was only a wooden building.   It had the largest floor in San Francisco, and was pressed into service at once.   The police and the troops, working in harmony together, passed the word that the dead and injured should be brought there, the hospitals and morgue having become choked, and the order was quickly obeyed, until about 400 of the hurt, many of them terribly mangled, were laid in improvised cots, attended by all the physicians and trained nurses who could be obtained.

The corpses were much fewer, the workers being too busy in fighting the fire and caring for the wounded to give time and attention as yet to the dead. But one of the first wagons to arrive brought a whole family—father, mother and three children—all dead except the baby, which had a broken arm and a terrible cut across the forehead. They had been dragged from the ruins of their house on the water front. A large consignment of bodies, mostly of workingmen, came from a small hotel on Eddy Street, through the roof of which the upper part of a tall building next door had fallen, crushing all below.

FIRE ATTACKS THE MINT.

To return to the story of the conflagration, the escape of the United States Mint was one of the most remarkable incidents. Within the vaults of this fine structure was the vast sum of $300,-000,000 in gold and silver coin and a value of $8,000,000 in bullion, and toward this mighty sum of wealth the flames swept on all sides, as if eager to add the reservoir of the precious metals to their spoils. The Mint building passed through the earthquake with little damage, though its big smokestacks were badly shaken. The fire seemed bent on making it its prey, every building around it being burned to the ground, and it remaining the only building for blocks that escaped destruction.

Its safety was due to the energy and activity of its employees. Superintendent Leach reached it shortly after the shock and found a number of men already there, whom he stationed at points of vantage from roof to basement. The fire apparatus of the Mint was brought into service and help given by the fire department, and after a period of strenuous labor the flames were driven back. The peril for a time was critical, the windows on Mint Avenue taking fire and also those on the rear three stories, and the flames for a time pouring in and driving back the workers. The roof also caught fire, but the men within fought like Titans, and efficient aid was given by a squad of soldiers sent to them. In the end the fire fiend was vanquished, though considerable damage was done to the adjusting rooms and the refinery, while the heavy stone cornice on that side of the building was destroyed. The total loss to the Mint was later estimated at $15,000.

Late on Wednesday evening the fire front crept close up to Mechanics' Pavilion, where a corps of fifty physicians and numerous nurses were active in the work of relief to the wounded. Ambulances and automobiles were busy unloading new patients rescued from the ruins when word came that the building would have to be vacated in haste. Every available vehicle was at once pressed into service and the patients removed as rapidly as possible, being taken to hospitals and private houses in the safer parts of the city. Hardly had the last of the injured been carried through the door when the roof was seen to be in a blaze, and shortly afterward the whole building burst into a whirlwind of flame.

At midnight the fire was raging and roaring with unslacked rage, and at dawn of Thursday its fury was undiminished. The work of destruction was already immense. In much of the Hayes

Valley district, south of McAllister and north of Market Street, the destruction was complete. From the Mechanics' Pavilion and St. Nicholas Hotel opposite down to Oakland Ferry the journey was heartrending, the scene appalling. On each side was ruin, nothing but ruin, and hillocks of masonry and heaps of rubbish of every description filled to its middle the city's greatest thoroughfare.

Across an alley from the Post Office stood the Grant Building, one of the headquarters of the army. Of this only the smoke-darkened walls were left. On Market Street opposite this building the beautiful front of the Hibernian Savings Bank, the favorite institution of the middle and poorer classes, presented a hideous aspect of ruin. At eleven o'clock of Wednesday night the north side of Market Street stood untouched, and hopes were entertained that the great Flood, Crocker, Phelan and other buildings would be spared, but the hunger of the fire fiend was not yet satiated, and the following day these proud structures had only their blackened ruins to show. On both sides of Market Street, down to the ferry, the tale was the same. The handsome and gigantic St. Francis Hotel, on Powell Street, fronting on Union Square, was left a ruined shell. This was one of the lofty steel structures that bore unharmed the earthquake shock, but quickly succumbed to the flames. Among the other skyscrapers north of Market Street that perished were the fourteen-story Merchants' Exchange, and the great Mills Building, occupying almost an entire block.

One section of the city that went without pity, as it had long stood with reprobation, was that group of disreputable buildings known as Chinatown, the place of residence of many thousands of Celestials. The flames made their way unchecked in this direction, and by noon on Thursday the whole section was a raging furnace.

the denizens escaping with what they could carry of their simple possessions. On the farther western side the flames cut a wide swath to Van Ness Avenue, a wide thoroughfare, at which it was hoped the march of the fire in this direction might be checked, especially as the water mains here furnished a weak supply.

In the Missouri district, to the south of Market Street, the zone of ruin extended westward toward the extreme southern portion, but was checked at Fourteenth and Missouri Streets by the whole-sale use of dynamite. At this point were located the Southern Pacific Hospital, the St. Francis Hospital and the College of Physicians and Surgeons. In order to save these institutions, buildings were blown up all around them, and by noon the danger was averted. It later became necessary to destroy the Southern Pacific Hospital with dynamite, the patients having been removed to places of safety.

THE PALACES ON NOB'S HILL.

In the centre of San Francisco rises the aristocratic elevation known as Nob's Hill, on which the early millionaires built their homes, and on which stood the city's most palatial residences. It ascends so abruptly from Kearney Street that it is inaccessible to any kind of vehicle, the slope being at any angle little short of forty-five degrees. It is as steep on the south side, and the only approach by carriage is from the north. To this hill is due the pioneer cable railway, built in the early '70's.

Here the "big four" of the railroad magnates—Stanford, Hopkins, Huntington and Crocker—had put millions in their mansions, the Mark Hopkins residence being said to have cost $2,500,-000. These men are all dead, and the last named edifice has been

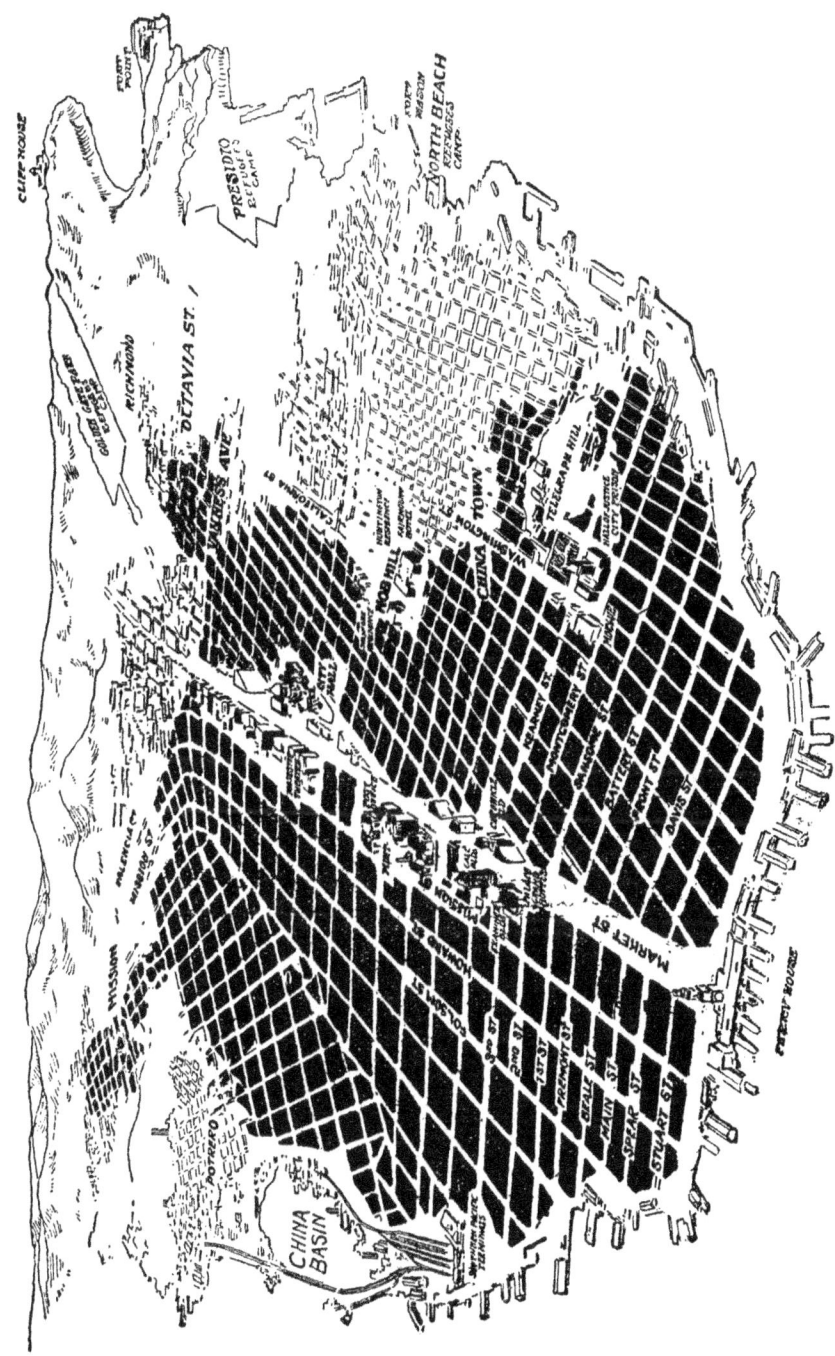

THE DEVASTATED AREA OF SAN FRANCISCO.
Bird'seye view of the portion destroyed by earthquake and fire.

47

converted into the Hopkins Art Institute, and at the time of the fire was well filled with costly art treasures. The Stanford Museum, which also contains valuable objects of art, is now the property of the Leland Stanford University. The Flood mansion, which cost more than $1,000,000, was one of the showy residences on this hill, west of it being the Huntington home and farther west the Crocker residence, with its broad lawns and magnificent stables. Many other beautiful and costly houses stood on this hill, and opposite the Stanford and Hopkins edifices the great Fairmount Hotel had for two years past been in process of construction and was practically completed. On the northeastern slope of this hill stood the famous Chinatown, through which it was necessary to pass to ascend Nob's Hill from the principal section of the wholesale district.

This region of palaces was the next to fall a prey to the insatiable flames. Early Thursday morning a change in the wind sent the fire westward, eating its way from the water front north of Market Street toward Nob's Hill. Steadily but surely it climbed the slope, and the Stanford and Hopkins edifices fell victims to its fury. Others of the palaces of millionairedom followed. Huge clouds of smoke enveloped the beautiful white stone Fairmount Hotel, and there was a general feeling of horror when this magnificent structure seemed doomed. To it the Committe of Safety had retreated, but the flames from the burning buildings opposite reached it, and the committee once more migrated in search of safe quarters. Fortunately, it escaped with little damage, its walls remaining intact and much of the interior being left in a state of preservation, warranting its managers to offer space within it to the committees whose aim it was to help the homeless or to store supplies. Some of the woodwork of the building was destroyed by the fire, but the structure was in such good condition that work on it was quickly resumed, with the statement that its completion would not be delayed more than three months beyond the date set, which was November, 1906.

In the district extending northwestwardly from Kearney Street and Montgomery Avenue, untouched during the first day, the fire spread freely on the second. This district embraces the Latin quarter, peopled by various nationalities, the houses being of the flimsiest construction. Once it had gained a foothold there, the fire swept onward as though making its way through a forest in the driest summer season.

An apochryphal incident is told of the fire in this quarter, which may be repeated as one example of the fables set afloat. It is stated that water to fight the fire here was sadly lacking, the only available supply being from an old well. At a critical moment the pump sucked dry, the water in the well being exhausted. The residents were not yet conquered. Some of them threw open their cellar doors and, calling for assistance, began to roll out barrels of red wine. Barrel after barrel appeared, until fully five hundred gallons were ready for use. Then the barrel heads were smashed in and the bucket brigade turned from water to wine. Sacks were dipped in the wine and used for fighting the fire. Beds were stripped of their blankets and these soaked in the wine and hung over exposed portions of the cottages, while men on the roofs drenched the shingles and sides of the houses with wine. The postscript to this queer story is that the wine won and the firefighters saved their homes. The story is worth retelling, though it may be added that wine, if it contained much alcohol, would serve as a feeder rather than as an extinguisher of flame.

A striking description of the aspect of the city on that terrible Wednesday is told by Jerome B. Clark, whose home was in Berkeley, but who did business in San Francisco. He left for the city early Wednesday morning, after a minor shake-up at home, which he thus describes:

A VIVID FIRE PICTURE.

"I was asleep and was awakened by the house rocking. With the exception of water in vases, and milk in pans being spilled, and one of our chimneys badly cracked, we escaped with nothing but a bad scare, but I can assure you it was a terrific and terrifying experience to feel that old house rocking, jolting and jumping under us, with the most terrible roar, dull, deep and nerve-racking. It calmed down after that and we went back to bed, only to get up at six o'clock to find that neighbors had suffered by having vases knocked from tables, bric-a-brac knocked around, tiles knocked out of grates and scarcely a chimney left standing. We thought that we had had the worst of it, so I started over to the city as usual, reaching there about eight o'clock, and it is just impossible to describe the scenes that met my eyes.

"In every direction from the ferry building flames were seething, and as I stood there, a five-story building half a block away fell with a crash, and the flames swept clear across Market Street and caught a new fireproof building recently erected. The streets in places had sunk three or four feet, in others great humps had appeared four or five feet high. The street car tracks were bent and twisted out of shape. Electric wires lay in every direction. Streets on all sides were filled with brick and mortar, buildings either completely collapsed or brick fronts had just dropped completely off. Wagons with horses hitched to them, drivers and all, lying on the streets, all dead, struck and killed by the falling bricks, these mostly the wagons of the produce dealers, who do the greater part of their work at that hour of the morning. Warehouses and large wholesale houses of all descriptions either down, or walls bulging, or else twisted, buildings moved bodily two or three feet out of a line and still standing with walls all cracked.

"The *Call* building, a twelve-story skyscraper, stood, and looked all right at first glance, but had moved at the base two feet at one end out into the sidewalk, and the elevators refused to work, all the interior being just twisted out of shape. It afterward burned as I watched it. I worked my way in from the ferry, climbing over piles of brick and mortar and keeping to the centre of the street and avoiding live wires that lay around on every side, trying to get to my office. I got within two blocks of it and was stopped by the police on account of falling walls. I saw that the block in which I was located was on fire, and seemed doomed, so turned back and went up into the city.

"Not knowing San Francisco, you would not know the various buildings, but fires were blazing in all directions, and all of the finest and best of the office and business buildings were either burning or surrounded. They pumped water from the bay, but the fire was soon too far away from the water front to make any efforts in this direction of much avail. The water mains had been broken by the earthquake, and so there was no supply for the fire engines and they were helpless. The only way out of it was to dynamite, and I saw some of the finest and most beautiful buildings in the city, new modern palaces, blown to atoms. First they blew up one or two

buildings at a time. Finding that of no avail, they took half a
block; that was no use; then they took a block; but in spite of them
all the fire kept on spreading.

"The City Hall, which, while old, was quite a magnificent building, occupying a large square block of land, was completely wrecked
by the earthquake, and to look upon reminded one of the pictures of
ancient ruins of Rome or Athens. The Palace Hotel stood for a long
time after everything near it had gone, but finally went up in smoke
as the rest. You could not look in any direction in the city but what
mass after mass of flame stared you in the face. To get about one
had to dodge from one street to another, back and forth in zigzag
fashion, and half an hour after going through a street, it would be
impassable. One after another of the magnificent business blocks
went down. The newer buildings seemed to have withstood the
shock better than any others, except well-built frame buildings.
The former lost some of the outside shell, but the frame stood all
right, and in some cases after fire had eaten them all to pieces, the
steel skeleton, although badly twisted and warped, still stood.

"When I finally left the city, it was all in flames as far as Eighth
Street, which is about a mile and a quarter or half from the water
front. I had to walk at least two miles around in order to get to
the ferry building, and when I got there you could see no buildings
standing in any direction. Nearly all the docks caved in or sheds
were knocked down, and all the streets along the water front were
a mass of seams, upheavals and depressions, car tracks twisted in
all shapes. Cars that had stood on sidings were all in ashes and still
burning."

Wednesday's conflagration continued unabated throughout
Thursday, and it was not until late on Friday that the fire-fighters

got it safely under control. They worked like heroes, struggling almost without rest, keeping up the nearly hopeless conflict until they fairly fell in their tracks from fatigue. Handicapped by the lack of water, they in one case brought it from the bay through lines of hose well on to a mile in length. Yet despite all they could do block after block of San Francisco's greatest buildings succumbed to the flames and sank in red ruin before their eyes.

THE LANDMARKS CONSUMED.

On all sides famous landmarks yielded to the fury of the flames. For three miles along the water front the ground was swept clean of buildings, the blackened beams and great skeletons of factories, warehouses and business edifices standing silhouetted against a background of flames, while the whole commercial and office quarter of Market Street suffered a similar fate. We may briefly instance some of these victims of the flames.

Among them were the Occidental Hotel, on Montgomery Street, for years the headquarters for army officers; the old Lick House, built by James Lick, the philanthropist; the California Hotel and Theatre, on Bush Street; and of theatres, the Orpheum, the Alcazar, the Majestic, the Columbia, the Magic, the Central, Fisher's and the Grand Opera House, on Missouri Street, where the Conried Opera Company had just opened for a two weeks' opera season.

The banks that fell were numerous, including the Nevada National Bank, the California, the Canadian Bank of Commerce, the First National, the London and San Francisco, the London, Paris and American, the Bank of British North America, the German-American Savings Bank and the Crocker-Woolworth Bank building. A large number of splendid apartment houses were also destroyed, and the tide of destruction swept away a host of noble buildings far too numerous to mention.

At Post Street and Grant Avenue stood the Bohemian Club, one of the widest known social organizations in the world. Its membership included many men famous in art, literature and commerce. Its rooms were decorated with the works of members, many of whose names are known wherever paintings are discussed and many of them priceless in their associations. Most of these were saved. There were on special exhibition in the "Jinks" room of the Bohemian Club a dozen paintings by old masters, including a Rembrandt, a Diaz, a Murillo and others, probably worth $100,000. These paintings were lost with the building, which went down in the flames.

One of the great losses was that of St. Ignatius' Church and College, at Van Ness Avenue and Hayes Street, the greatest Jesuitical institution in the west, which cost a couple of millions of dollars. The Merchants' Exchange building, a twelve-story structure, eleven of whose floors were occupied as offices by the Southern Pacific Railroad Company, was added to the sum of losses.

THE FIRE UNDER CONTROL.

For three long days the terrible fire fiend kept up his work, and the fight went on until late on Friday, when the sweep of the flames was at length checked and the fire brought under control. The principal agent in this victory was dynamite, which was freely used. To its work a separate chapter will be devoted. When at length the area of the conflagration was limited the wealthiest part of the city lay in embers and ashes, one of the principal localities to escape being Pacific Heights, a mile west from Nob's Hill, on which stood many costly homes of recent construction.

On Friday night the fire that had worked its way from Nob's Hill to North Beach Street, sweeping that quarter clean of buildings, veered before a fierce wind and made its way southerly to the great sea wall, with its docks and grain warehouses. The flames reached the tanks of the San Francisco Gas Company, which had previously been pumped out, and on Saturday morning the grain sheds on the water front, about half a mile north of the ferry station, were fiercely burning. But the fire here was confined to a small area, and, with the work of fireboats in the bay and of the firemen on shore, who used salt water pumped into their engines, it was prevented from reaching the ferry building and the docks in that vicinity.

The buildings on a high slope between Van Ness and Polk Streets, Union and Filbert Streets, were blazing fiercely, fanned by a high wind, but the blocks here were so thinly settled that the fire had little chance of spreading widely from this point. In fact, it was at length practically under control, and the entire western addition of the city west of Van Ness Avenue was safe from the flames. The great struggle was fairly at an end, and the brave force of workers were at length given some respite from their strenuous labors.

During the height of the struggle and the days of exhaustion and depression that followed, exaggerated accounts of the losses and of the area swept by the flames were current, some estimate making the extent of the fire fifteen square miles out of the total of twenty-five square miles of the city's area. It was not until Friday, the 27th, that an official survey of the burned district, made by City Surveyor

Woodward, was completed, and the total area burned over found to be 2,500 acres, a trifle less than four square miles. This, however, embraced the heart of the business section and many of the principal residence streets, much of the saved area being occupied by the dwellings of the poorer people, so that the money loss was immensely greater than the percentage of ground burned over would indicate.

THE BURNING CITY OF SAN FRANCISCO.

Photographed from a boat on San Francisco Bay on Thursday, April 19th, the day after the earthquake, showing the conflagration at its height. The two pillars of smoke stood over the doomed city for two days. The ferryboat in the right foreground is loaded with refugees rushing to Oakland.

A TERRIBLE BUT MAGNIFICENT SPECTACLE PHOTOGRAPHED IN THE
CENTRE OF SAN FRANCISCO.

Panorama of the heart of the business district, in which all the buildings were destroyed or
badly damaged by the earthquake and the fire.

IN THE FIRE-SWEPT TENEMENT DISTRICT OF SAN FRANCISCO.

This photograph was made immediately after the flames had devoured what the earthquake left of the homes of the poor. The flames are seen in the background, moving on to consume the taller buildings whose outlines appear dimly through the dense smoke.

A BURNING STREET IN SAN FRANCISCO.

A photograph of Front Street, between Howard and Market, showing the extensive and flourishing manufacturing district all ablaze. The damage done by the earthquake is visible at the tops of the buildings in the foreground.

CHAPTER III.

Fighting the Flames With Dynamite.

SHAKEN by earthquake, swept by flames, the water supply cut off by the breaking of the mains, the authorities of the doomed city for a time stood appalled. What could be done to stay the fierce march of the flames which were sweeping resistlessly over palace and hovel alike, over stately hall and miserable hut? Water was not to be had; what was to take its place? Nothing remained but to meet ruin with ruin, to make a desert in the path of the fire and thus seek to stop its march. They had dynamite, gunpowder and other explosives, and in the frightful exigency there was nothing else to be used. Only for a brief interval did the authorities yield to the general feeling of helplessness. Then they aroused themselves to the demands of the occasion and prepared to do all in the power of man in the effort to arrest the conflagration.

While the soldiers under General Funston took military charge of the city, squads of cavalry and troops of infantry patrolling the streets and guarding the sections that had not yet been touched by the flames, Mayor Schmitz and Chief of Police Dinan sprang into the breach and prepared to make a desperate charge against the platoons of the fire. This was not all that was needed to be done. From the "Barbary Coast," as the resort of the vicious and criminal classes was called, hordes of wretches poured out as soon as night fell, seeking to slip through the guards and loot stores and rob the dead in the burning section. Orders were given to the soldiers to kill all who were engaged in such work, and these orders were carried out. An associated Press reporter saw three of these thieves shot and fatally wounded, and doubtless others of them were similarly dealt with elsewhere.

A band of fire-fighters was quickly organized by the Mayor and Chief of Police, and the devoted firemen put themselves in the face of the flames, determined to do their utmost to stay them in their course. Cut off from the use of their accustomed engines and water streams, which might have been effective if brought into play at the beginning of the struggle, there was nothing to work with but the dynamite cartridge and the gunpowder mine, and they set bravely to work to do what they could with these. On every side the roar of explosions could be heard, and the crash of falling walls came to the ear, while people were forced to leave buildings which still stood, but which it was decided must be felled. Frequently a crash of stone and brick, followed by a cloud of dust, gave warning to pedestrians that destruction was going on in the forefront of the flames, and that travel in such localities was unsafe.

FIGHTING THE FLAMES.

All through the night of Wednesday and the morning of Thursday this work went on, hopelessly but resolutely. During the following day blasts could be heard in different sections at intervals of a few minutes, and buildings not destroyed by fire were blown to atoms, but over the gaps jumped the live flames, and the disheartened fire-fighters were driven back step by step; but they continued the work with little regard for their own safety and with unflinching desperation.

One instance of the peril they ran may be given. Lieutenant Charles O. Pulis, commanding the Twenty-fourth Company of Light Artillery, had placed a heavy charge of dynamite in a building at Sixth and Jesse Streets. For some reason it did not explode, and he returned to relight the fuse, thinking it had become extinguished. While he was in the building the explosion took place, and he received injuries that seemed likely to prove fatal, his skull being fractured and several bones broken, while he was injured internally. In the early morning, when the fire reached the municipal building on Portsmouth Square, the nurses, with the aid of soldiers, got out fifty bodies which were in the temporary morgue and a number of patients from the receiving hospital. Just after they reached the street with their gruesome charge a building was blown up, and the flying bricks and splinters came falling upon them. The nurses fortunately escaped harm, but several of the soldiers were hurt, and had to be taken with the other patients to the out-of-doors Presidio hospital.

The Southern Pacific Hospital, at Fourteenth and Missouri Streets, was among the buildings destroyed by dynamite, the patients having been removed to places of safety, and the Linda Vista and the Pleasanton, two large family hotels on Jones Street, in the better part of the city, were also among those blown up to stay the progress of the conflagration.

THE STRUGGLE AGAINST THE FIRE.

The fire had continued to creep onward and upward until it reached the summit of Nob Hill, a district of splendid residences, and threatened the handsome Fairmount Hotel, then the headquarters of the Municipal Council, acting as a Committee of Public Safety.

As day broke the flames seized upon this beautiful structure, and the Council was forced to retreat to new quarters. They finally met in the North End Police Station, on Sacramento Street, and there entered actively upon their duties of seeking to check the progress of the flames, maintain order in the city and control and direct the host of fugitives, many of whom, still in a state of semi-panic, were moving helplessly to and fro and sadly needed wise counsels and a helping hand.

The fire-fighters meanwhile kept up their indefatigable work under the direction of the Mayor and the chief of their department. The engines almost from the start had proved useless from lack of water, and were either abandoned or moved to the outlying districts, in the vain hope that the water mains might be repaired in time to permit of a final stand against the whirlwind march of the flames. The cloud of despair grew darker still as the report spread that the city's supply of dynamite had given out.

"No more dynamite! No more dynamite!" screamed a fireman as he ran up Ellis Street past the doomed Flood building at two o'clock on Friday morning, tears standing in his smoke-smirched eyes.

"No more dynamite! O God! no more dynamite! We are lost!" moaned the throng that heard his despairing words.

A NEW SUPPLY OF EXPLOSIVES.

So, at that hour, the supply of the explosive exhausted, and not a dozen streams of water being thrown in the entire fire zone, the stunned firemen and the stupefied people stood helpless with their eyes fixed in despair upon the swiftly creeping flames.

Had all been like these the entire city would have been doomed, but there were those at the head of affairs who never for a moment gave up their resolution. Dynamite and giant powder were to be had in the Presidio military reservation, and a requisition upon the army authorities was made. The louder reverberations as the day advanced and night came on showed that a fresh supply had been obtained, and that a new and determined campaign against the conflagration had been entered upon. Hitherto much of the work had been ignorantly and carelessly done, and by the hasty and premature use of explosives more harm than good had been occasioned.

As the fire continued to spread in spite of the heroic work of the fighting corps, the Committee of Safety called a meeting at noon on Friday and decided to blow up all the residences on the east side of Van Ness Avenue, between Golden Gate and Pacific Avenues, a distance of one mile. Van Ness Avenue is one of the most fashionable streets of the city and has a width of 125 feet, a fact which led to the idea that a safety line might be made here too broad for the flames to cross.

The firemen, therefore, although exhausted from over twenty-four hours' work and lack of food, determined to make a desperate stand at this point. They declared that should the fire cross Van Ness Avenue and the wind continue its earlier direction toward the west, the destruction of San Francisco would be virtually complete. The district west of Van Ness Avenue and north of McAllister constitutes the finest part of the metropolis. Here are located all of the finer homes of the well-to-do and wealthier classes, and the resolution to destroy them was the last resort of desperation.

Hundreds of police, regiments of soldiers and scores of volunteers were sent into the doomed district to warn the people to flee. They heroically responded to the demand of law and went bravely on their way, leaving their loved homes and trudging painfully over the pavements with the little they could carry away of their treasured possessions.

The reply of a grizzled fire engineer standing at O'Farrell Street and Van Ness Avenue, beside a blackened engine, may not have been as terse as that of Hugo's guardsman at Waterloo, but the pathos of it must have been as great. In answer to the question of what they proposed to do, he said:

"We are waiting for it to come. When it gets here we will make one more stand. If it crosses Van Ness Avenue the city is gone."

THE SAVERS OF THE CITY.

Yet the work now to be done was much too important to be left to the hands of untrained volunteers. Skilled engineers were needed, men used to the scientific handling of explosives, and it was men of this kind who finally saved what is left to-day of the city. Three men saved San Francisco, so far as any San Francisco existed after the fire had worked its will, these three constituting the dynamite squad who faced and defied the demon at Van Ness Avenue.

When the burning city seemed doomed and the flames lit the sky farther and farther to the west, Admiral McCalla sent a trio of his most trusted men from Mare Island with orders to check the conflagration at any cost of property. With them they brought a ton and a half of guncotton. The terrific power of the explosive was equal to the maniac determination of the fire. Captain Mac-

Bride was in charge of the squad, Chief Gunner Adamson placed the charges and the third gunner set them off.

Stationing themselves on Van Ness Avenue, which the conflagration was approaching with leaps and bounds from the burning business section of the city, they went systematically to work, and when they had ended a broad open space, occupied only by the dismantled ruins of buildings, remained of what had been a long row of handsome and costly residences, which, with all their treasures of furniture and articles of decoration, had been consigned to hideous ruin.

The thunderous detonations, to which the terrified city listened all that dreadful Friday night, meant much to those whose ears were deafened by them. A million dollars' worth of property, noble residences and worthless shacks alike, were blown to drifting dust, but that destruction broke the fire and sent the raging flames back over their own charred path. The whole east side of Van Ness Avenue, from the Golden Gate to Greenwich, a distance of twenty-two blocks, or a mile and a half, was dynamited a block deep, though most of the structures as yet had stood untouched by spark or cinder. Not one charge failed. Not one building stood upon its foundation.

Unless some second malicious miracle of nature should reverse the direction of the west wind, by nine o'clock it was felt that the populous district to the west, blocked with fleeing refugees and unilluminated except by the disastrous glare on the water front, was safe. Every pound of guncotton did its work, and though the ruins burned, it was but feebly. From Golden Gate Avenue north the fire crossed the wide street in but one place. That was at the Claus Spreckels place, on the corner of California Street.

There the flames were writhing up the walls before the dyna-
miters could reach the spot.   Yet they made their way to the foundations, carrying their explosives, despite the furnace-like heat.   The
charge had to be placed so swiftly and the fuse lit in such a hurry
that the explosion was not quite successful from the trained view-
point of the gunners.   But though the walls still stood, it was only
an empty victory for the fire, as bare brick and smoking ruins are
poor food for flames.

Captain MacBride's dynamiting squad had realized that a
stand was hopeless except on Van Ness Avenue, their decision thus
coinciding with that of the authorities.   They could have forced
their explosives farther in the burning section, but not a pound of
guncotton could be or was wasted.   The ruined blocks of the wide
thoroughfare formed a trench through the clustered structures that
the conflagration, wild as it was, could not leap.   Engines pumping
brine through Fort Mason from the bay completed the little work
that the guncotton had left, but for three days the haggard-eyed
firemen guarded the flickering ruins.

The desolate waste straight through the heart of the city
remained a mute witness to the most heroic and effective work of
the whole calamity.   Three men did this, and when their work was
over and what stood of the city rested quietly for the first time,
they departed as modestly as they had come.   They were ordered
to save San Francisco, and they obeyed orders, and Captain Mac-
Bride and his two gunners made history on that dreadful night.

They stayed the march of the conflagration at that critical
point, leaving it no channel to spread except along the wharf region,
in which its final force was spent.   One side of Van Ness Avenue
was gone; the other remained, the fire leaping the broad open space
only feebly in a few places, where it was easily extinguished.

In this connection it is well to put on record an interesting circumstance. This is that there is one place within pistol shot of San Francisco that the earthquake did not touch, that did not lose a chimney or feel a tremor. That spot is Alcatraz Island. Despite the fact that the island is covered with brick buildings, brick forts and brick chimneys, not a brick was loosened nor a crack made nor a quiver felt. When the scientist comes to write he will have his hands full explaining why Alcatraz did not have any physical knowledge of the event. It was as if New York were to be shaken to its foundation, and Governor's Island, quietly pursuing its military routine, should escape without a qualm.

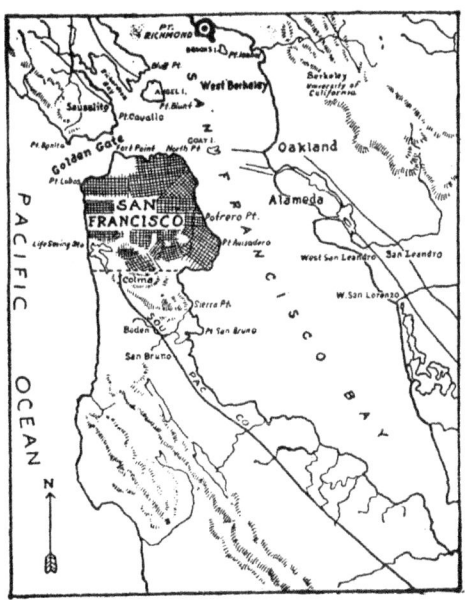

CHAPTER IV.

The Reign of Destruction and Devastation

RARELY, in the whole history of mankind, has a great city been overwhelmed by destruction so suddenly and awfully as was San Francisco. One minute its inhabitants slept in seeming safety and security. Another minute passed and the whole great city seemed tumbling around them, while sights of terror met the eyes of the awakened multitude and sounds of horror came to their ears. The roar of destruction filled the air as the solid crust of the earth lifted and fell and the rocks rose and sank in billowing waves like those of the open sea.

Not all, it is true, were asleep. There was the corps of night workers, whose duties keep them abroad till day dawns. There were those whose work calls them from their homes in the early morn. People of this kind were in the streets and saw the advent of the reign of devastation in its full extent. From the story of one of these, P. Barrett, an editor on the *Examiner*, we select a thrilling account of his experience on that morning of awe.

AN EDITOR'S NARRATIVE.

"I have seen this whole, great horror. I stood with two other members of the *Examiner* staff on the corner of Market Street, waiting for a car. Newspaper duties had kept us working until five o'clock in the morning. Sunlight was coming out of the early morning mist. It spread its brightness on the roofs of the sky-scrapers, on the domes and spires of churches, and blazed along up the wide street with its countless banks and stores, its restaurants and cafes. In the early morning the city was almost noiseless. Occasionally a newspaper wagon clattered up the street or a milk wagon rumbled along. One of my companions had told a funny story. We were laughing at it. We stopped—the laugh unfinished on our lips.

"Of a sudden we had found ourselves staggering and reeling. It was as if the earth was slipping gently from under our feet. Then came a sickening swaying of the earth that threw us flat upon our faces. We struggled in the street. We could not get on our feet.

"I looked in a dazed fashion around me. I saw for an instant the big buildings in what looked like a crazy dance. Then it seemed as though my head were split with the roar that crashed into my ears. Big buildings were crumbling as one might crush a biscuit in one's hand. Great gray clouds of dust shot up with flying timbers, and storms of masonry rained into the street. Wild, high jangles of smashing glass cut a sharp note into the frightful roaring. Ahead of me a great cornice crushed a man as if he were a maggot—a laborer in overalls on his way to the Union Iron Works, with a dinner pail on his arm.

"Everywhere men were on all fours in the street, like crawling bugs. Still the sickening, dreadful swaying of the earth continued. It seemed a quarter of an hour before it stopped. As a matter of fact, it lasted about three minutes. Footing grew firm again, but hardly were we on our feet before we were sent reeling again by repeated shocks, but they were milder. Clinging to something, one could stand.

"The dust clouds were gone. It was quite dark, like twilight. But I saw trolley tracks uprooted, twisted fantastically. I saw wide wounds in the street. Water flooded out of one. A deadly odor of gas from a broken main swept out of the other. Telegraph poles were rocked like matches. A wild tangle of wires was in the street. Some of the wires wriggled and shot blue sparks.

"From the south of us, faint, but all too clear, came a horrible chorus of human cries of agony. Down there in a ramshackle section of the city the wretched houses had fallen in upon the sleeping families. Down there throughout the day a fire burned the great part of whose fuel it is too gruesome a thing to contemplate.

"That was what came next—the fire. It shot up everywhere. The fierce wave of destruction had carried a flaming torch with it —agony, death and a flaming torch. It was just as if some fire demon was rushing from place to place with such a torch."

WRECK AND RUIN.

The magnitude of the calamity became fully apparent after the sun had risen and began to shine warmly and brightly from the east over the ruined city. Old Sol, who had risen and looked down upon this city for thousands of times, had never before seen such a spectacle as that of this fateful morning. Where once rose noble buildings were now to be seen cracked and tottering walls, fallen chimneys, here and there fallen heaps of brick and mortar, and out of and above all the red light of the mounting flames. From the middle of the city's greatest thoroughfare ruin, only ruin, was to be seen on all sides. To the south, in hundreds of blocks, hardly a

A CALIFORNIA VILLA AT MONTECITO.

Many of these beautiful homes were destroyed by the earthquake and afterwards consumed by fire.

building had escaped unscathed. The cracked walls of the new Post Office showed the rending power of the earthquake. A part of the splendid and costly City Hall collapsed, the roof falling to the court-yard and the smaller towers tumbling down. Some of the wharves, laden with goods of every sort, slid into the bay. With them went thousands of tons of coal. On the harbor front the earth sank from six to eight inches, and great cracks opened in the streets.

San Francisco's famous Chinatown, the greatest settlement of the Celestials on this continent, went down like a house of cards. When the earthquake had passed this den of squalor and infamy was no more. The Chinese theatres and joss-houses tumbled into ruins, rookery after rookery collapsed, and hundreds of their inhabitants were buried alive. Panic reigned supreme among the fugitives, who filled the streets in frightened multitudes, dragging from the wreck whatever they could save of their treasured possessions. Much the same was the case with the Japanese quarter, which fire quickly invaded, the people fleeing in terror, carrying on their backs what few of their household effects they were able to rescue.

As for the people of Chinatown, however, no one knows or will ever know the extent of the dread fate that overcame them, for no one knows the secrets of that dark abode of infamy and crime, whose inhabitants burrowed underground like so many ants, and hid their secrets deep in the earth.

THE RUIN OF CHINATOWN.

W. W. Overton, of Los Angeles, thus describes the Chinatown dens and the revelations made by the earthquake and the flames:

"Strange is the scene where San Francisco's Chinatown stood.

No heap of smoking ruins marks the site of the wooden warrens where the Orientals dwelt in thousands. Only a cavern remains, pitted with deep holes and lined with dark passageways, from whose depths come smoke wreaths. White men never knew the depth of Chinatown's underground city. Many had gone beneath the street level two and three stories, but now that the place had been unmasked, men may see where its inner secrets lay. In places one can see passages a hundred feet deep.

"The fire swept this Mongolian quarter clean. It left no shred of the painted wooden fabric. It ate down to the bare ground, and this lies stark, for the breezes have taken away the light ashes. Joss houses and mission schools, groceries and opium dens, gambling resorts and theatres, all of them went. These buildings blazed up like tissue paper.

"From this place I saw hundreds of crazed yellow men flee. In their arms they bore opium pipes, money bags, silks and children. Beside them ran the trousered women and some hobbled painfully. These were the men and women of the surface. Far beneath the street levels in those cellars and passageways were other lives. Women, who never saw the day from their darkened prisons, and their blinking jailors were caught and eaten by the flames."

Devastation spread widely on all sides, ruining the homes of the rich as well as of the poor, of Americans as well as of Europeans and Asiatics, the marts of trade, the haunts of pleasure, the realms of science and art, the resorts of thousands of the gay population of the Golden State metropolis. To attempt to tell the whole story of destruction and ruin would be to describe all for which San Francisco stood. Science suffered in the loss of the San Francisco Academy of Sciences, which was destroyed with its invaluable contents. This building, erected fifteen years ago at a cost of $500,000, was a seven-story building with a rich collection of objects of science. Much of the academy's contents can never be replaced. It represented the work of many years. There was a rare collection of Pacific Sea birds which was the most valuable of its kind in the world. In fact, the entire collection of birds ranked very high, was visited by ornithologists from every country, and was the pride of the city. The academy was founded in 1850, James Lick, the same man who endowed the Lick Observatory, giving it $1,000,000, so it was on a prosperous footing. It will take many years of active labor to replace the losses of an hour or two of the reign of fire in this institution, while much that it held is gone beyond restoration.

LOSS TO ART AND SCIENCE.

Art suffered as severely as science, the valuable collections in private and public buildings being nearly all destroyed. We have spoken of the rare paintings burned in the Bohemian Club building. The collections on Nob's Hill suffered as severely. When the mansions here, the Fairmount Hotel and Mark Hopkins Institute were approached by the flames, many attempts were made to remove some of the priceless works of art from the buildings. A crowd of soldiers was sent to the Flood and the Huntington mansions and the Hopkins Institute to rescue the paintings. From the Huntington home and the Flood mansion canvases were cut from the framework with knives. The collections in the three buildings, valued in the hundreds of thousands, in great part were destroyed, few being saved from the ravages of the fire..

The destruction of the libraries, with their valuable collections of books, was also a very serious loss to the city and its people. Of these there were nine of some prominence, the Sutro Library containing many rare books among its 200,000 volumes, while that of the Mechanics Institute possessed property valued at $2,000,000. The Public Library occupied a part of the City Hall, the new building proposed by the city, with aid to the extent of $750,000 by Andrew Carnegie, being fortunately still in embryo.

In the burning of the banks the losses were limited to the buildings, their money and other valuables being securely locked in fireproof vaults. But these became so heated by the flames that it was necessary to leave them to a gradual cooling for days, during which their treasures were unavailable, and those with deposits, small or large, were obliged to depend on the benevolence of the nation for food, such wealth as was left to them being locked up beyond their reach. It was the same with the United States Sub-Treasury, which was entirely destroyed by fire, its vaults, which contained all the cash on hand, being alone preserved. Guards were put over these to protect their contents against possible loss by theft.

One serious effect of the conflagration was the general disorganization of the telegraph system. News items were sent over the wires, but private messages inquiring about missing friends for days failed to reach the parties concerned or to bring any return.

That the world received news of the San Francisco disaster during the dread day after the earthquake is due in part to the courage of the telegraph operators, who stuck to their posts and continued to send news and other messages in spite of great personal danger.

The operators and officials of the Postal Telegraph Company remained in the main office of the company, at the corner of Market and Montgomery Streets, opposite the Palace Hotel, until they were ordered out of it because of the danger of the dynamite explosions in the immediate vicinity. The men proceeded to Oakland, across the bay, and took possession of the office there. That night the company operated seven wires from Oakland, all messages from the city being taken across the bay in boats. As the days passed on the service gradually improved, but a week or more passed away before the general service of the company became satisfactory.

THE DANGER FROM THIRST.

Such news as came from the city was full of tales of horror. For a number of days one of the chief sources of trouble was from thirst. Although the earthquake shocks had broken water mains in probably hundreds of places, strange to say, no water, or very little at least, appeared on the surface of the ground. Public fountains on Market. Street gave out no relief to the thirsty thousands. At Powell and Market Streets a small stream of water spurted up through the cobblestones and formed a muddy pool, at which the thirsty were glad enough to drink. The soldiers, disregarding the order not to let people move about, permitted bucket brigades to go forth and bring back water to relieve the women and the crying children. To reach the water it was necessary sometimes to go a mile to one of the four reservoirs which top the hills.

Here is a story told by one observer of incidents in the city during the fire:

"I talked to one man who slept in Alta Plaza. The fire was

going on in the district south of them, and at intervals all night exhausted fire-fighters made their way to the plaza and dropped, with the breath out of them, among the huddled people and the bundles of household goods. The soldiers, who are administering affairs with all the justice of judges and all the devotion of heroes, kept three or four buckets of water, even from the women, for these men, who kept coming all night long. There was a little food, also kept by the soldiers for these emergencies, and the sergeant had in his charge one precious bottle of whisky, from which he doled out drinks to those who were utterly exhausted.

"Over in a corner of the plaza a band of men and women were praying, and one fanatic, driven crazy by horror, was crying out at the top of his voice:

" 'The Lord sent it, the Lord!'

"His hysterical crying got in the nerves of the soldiers and bade fair to start a panic among the women and children, so the sergeant went over and stopped it by force. All night they huddled together in this hell, with the fire making it bright as day on all sides; and in the morning the soldiers, using their sense again, commandeered a supply of bread from a bakery, sent out another water squad, and fed the refugees with a semblance of breakfast.

"There was one woman in the crowd who had been separated from her husband in a rush of the smoke and did not know whether he was living. The women attended to her all night and in the morning the soldiers passed her through the lines in her search. A few Chinese made their way into the crowd. They were trembling, pitifully scared and willing to stop wherever the soldiers placed them. This is only a glimpse of the horrible night in the parks and open places.

"We learn here that many of the well-to-do people in the upper residence district have gathered in the strangers from the highways and byways and given them shelter and comfort for the night in their living rooms and drawing rooms. Shelter seems to have come more easily than food. Not an ounce of supplies, of course, has come in for two days, and most of the permanent stores are in the hands of the soldiers, who dole them out to all comers alike. But the hungry cannot always find the military stores and the news has not gotten about, since there are no newspapers and no regular means of communication.

"An Italian tells me that he was taken in by a family living in a three-story house in the fashionable Pacific Avenue. There were twenty refugees who passed the night in the drawing room of that house, whose mistress took down hangings to make them comfortable. In the morning all the food that was left over in that home of wealth was enough flour and baking powder to shake together a breakfast for the refugees. They were hardly ready to leave that house when the fire came their way, and the people of the house, together with the refugees, who included two Chinese, made their way to the open ground of the Presidio. With them streamed a procession of folks carrying valuables in bundles.

"There came out, too, tales of both heroism and crime. The firemen had been at it for thirty-six hours under such conditions as firemen never before faced, and they do little more than give directions, while the volunteers, thousands of young Western men who have remained to see it through, do the work. The troops have all that they can do to handle the crowds in the streets and prevent panics. The work of dynamiting, tearing down and rescuing is in the hands of the volunteers.

"This morning an eddy of flame from the edge of the burning wholesale district ran up the slope of Russian Hill, the highest eminence in the city. All along the edge of that hill and up the slopes are little frame houses which hold Italians and Mexicans. A corps of volunteer aides ran along the edge of the fire, warning people out of the houses. But the flames ran too fast and three women were caught in the upper story of an old frame house. A young man tore a rail from a fence, managed to climb it, and reached the window. He bundled one woman out and slid her down the rail; then the roof caught fire. He seized another woman and managed to drop her on the rail, down which she slid without hurting herself a great deal. But the roof fell while he was struggling with another woman and they fell together into the flames. There must have been hundreds of such heroisms and dozens of such catastrophes. We are so drunken and dulled by horror that we take such stories calmly now. We are saturated."

HOW LOOTING WAS HINDERED.

One thing to be strictly guarded against in those days of destruction was the outbreak of lawlessness. A city as large as San Francisco is sure to hold a large number of the brigands of civilization, a horde who need to be kept under strict discipline at all times, and especially when calamity lets down for the time being the bars of the law, at which time many of the usually law-abiding would join their ranks if any license were allowed. The authorities made haste to guard against this and certain other dangers, Mayor Schmitz issuing on Wednesday the following proclamation:

"The Federal troops, the members of the regular police force and special police officers have been authorized to kill any and all persons engaged in looting or in the commission of any other crime.

"I have directed all the gas and electric lighting companies not to turn on gas or electricity until I order them to do so. You may, therefore, expect the city to remain in darkness for an indefinite time.

"I request all citizens to remain at home from darkness until daylight every night until order is restored.

"I warn all citizens of the danger of fire from damaged or destroyed chimneys, broken or leaking gas pipes or fixtures or any like causes."

He also ordered that no lights should be used in the houses and no fires built in the houses until the chimneys had been inspected and repaired.

There was need of vigilance in this direction, for the vandals were quickly at work. Routed out from their dens along the wharves, the rats of the waterfront, the drifters on the back eddy of civilization, crawled out intent on plunder. Early in the day a policeman caught one of these men creeping through the window of a small bank on Montgomery Street and shot him dead. But the police were kept too busy at other necessary duties to devote much time to these wretches, and for a time many of them plundered at will, though some of them met with quick and sure retribution.

STORIES BY SIGHTSEERS.

One onlooker says: "Were it not for the fact that the soldiers in charge of the city do not hesitate in shooting down the ghouls the lawless element would predominate. Not alone do the soldiers execute the law. On Wednesday afternoon, in front of the Palace Hotel, a crowd of workers in the mines discovered a miscreant in the act of robbing a corpse of its jewels. Without delay he was seized, a rope obtained, and he was strung up to a beam that was left standing in the ruined entrance of the hotel. No sooner had he been hoisted up and a hitch taken in the rope than one of his fellow-criminals was captured. Stopping only to obtain a few yards of hemp, a knot was quickly tied, and the wretch was soon adorning the hotel entrance by the side of the other dastard.

"These are the only two instances I saw, but I heard of many that were seen by others. The soldiers do all they can, and while the unspeakable crime of robbing the dead is undoubtedly being practiced, it would be many times as prevalent were it not for the constant vigilance on all sides, as well as the summary justice."

Another observer tells of an instance of this summary justice that came under his eyes:

"At the corner of Market and Third Streets on Wednesday I saw a man attempting to cut the fingers from the hand of a dead woman in order to secure the rings which adorned the stiffened fingers. Three soldiers witnessed the deed at the same time and ordered the man to throw up his hands. Instead of obeying the command he drew a revolver from his pocket and began to fire at his pursuer without warning. The three soldiers, reinforced by half a dozen uniformed patrolmen, raised their rifles to their shoulders and fired. With the first shots the man fell, and when the soldiers went to the body to dump it into an alley nine bullets were found to have entered it."

The warning this severity gave was accentuated in one instance in a most effective manner. On a pile of bricks, stones and rubbish was thrown the body of a man shot through the heart, and on his chest was pinned this placard:

"Take warning!"

Those of the ghouls who saw this were likely to desist from their detestable work, unless they valued spoils more than life.

Willis Ames, a Salt Lake City man, tells of the kind of justice done to thieves, as it came under his observation:

"I saw man after man shot down by the troops. Most of these were ghouls. One man made the trooper believe that one of the dead bodies lying on a pile of rocks was his mother, and he was permitted to go up to the body. Apparently overcome by grief, he threw himself across the corpse. In another instant the soldiers discovered that he was chewing the diamond earrings from the ears of the dead woman. 'Here is where you get what is coming to you,' said one of the soldiers, and with that he put a bullet through the ghoul. The diamonds were found in the man's mouth afterward."

Others were shot to save them from the horror of being burned alive. Max Fast, a garment worker, tells of such an instance. He says:

"When the fire caught the Windsor Hotel at Fifth and Market Streets there were three men on the roof, and it was impossible to get them down. Rather than see the crazed men fall in with the roof and be roasted alive the military officer directed his men to shoot them, which they did in the presence of 5,000 people."

He further states: "At Jefferson Square I saw a fatal clash between the military and the police. A policeman ordered a soldier to take up a dead body to put it in the wagon, and the soldier ordered the policeman to do it. Words followed, and the soldier shot the policeman dead."

Among the many stories of this character on record is that of a concerted effort to break into and rob the Mint, which led to the death of fourteen men, who were shot down by the guard in charge. They had disregarded the command of the officer in charge to desist. They disobeyed, and the death of nearly the whole of them followed.

DEATH FOR SLIGHT OFFENSE.

As may well be imagined, the privilege given to fire at will was very likely to lead to examples of unjustifiable haste in the use of the rifle. Such haste is not charged against the United States troops, but the militia and volunteer guards showed less judgment in the use of their weapons. Thus we are told that one man was shot for the minor offense of washing his hands in drinking water which had been brought with great trouble for the thirsty people gathered in Columbia Park. It is also said that a bank clerk, searching the ruins of his bank under orders, was killed by a soldier who thought he was looting. More than one seems to have been shot as looters for entering their own homes.

Among the reports there is one that two men were shot through the windows of their houses because they disobeyed the general orders and lit candles, and one woman because she lighted a fire in her cook stove. Yet, if such unwarranted acts existed, there were others better deserved. It is said that three men were lined up and shot before ten thousand people. One was caught taking the rings from a woman who had fainted, another had stolen a piece of bread from a hungry child, and the third, little more than a boy, was found in the act of robbing tents. One thief who escaped the bullet richly deserved it. He came upon a Miss Logan when lying unconscious on the floor of the St. Francis Hotel after the earthquake, and, rather than take the time to wrench some valuable rings from her hand, cut off the finger bearing them, and left her to the horrors of the coming fire.

The climax in the too free use of the rifle came on the 23d, when Major H. C. Tilden, a prominent member of the General Relief Committee, was shot and killed in his automobile by members of the citizens' patrol. Two others in the car were struck by bullets. The automobile had been used as an ambulance and the Red Cross flag was displayed on it. The excuse of the shooters was that they did not see the flag and that the car did not stop when challenged. This act led to an order forbidding the carrying of firearms by the citizens' committees and to stricter regulation of the soldiers in the use of their weapons.

Later on looting took a new form different from that at first shown and was practiced by a different class of people. These were the sightseers, many of them people of prominence, who entered upon a crusade of relic hunting in Chinatown, gathering and carrying off from the ashes of this quarter valuable pieces of chinaware, bronze ornaments, etc. It became necessary to put a stop to this, and on April 30th four militiamen were arrested while digging in the ruins of the Chinese bazaars, and others were frightened away by shots fired over their heads. A strong military line was then drawn around the district, and this last resource of the looter came to an end.

CHAPTER V.

The Panic Flight of a Homeless Host.

THE scene that was visible in the streets of San Francisco on that dread Wednesday morning was one to make the strongest shudder with horror. Those three minutes of devastating earth tremors were moments never to be forgotten. In such a time it is the human instinct to get into the open air, and the people stumbled from their heaving and quivering houses to find even the solid earth was swaying and rising and falling, so that here and there great rents opened in the streets. To the panic-stricken people the minutes that followed seemed years of terror. Doubtless some among them died of sheer fright and more went mad with terror. There was a roar in the air like a burst of thunder, and from all directions came the crash of falling walls. They would run forward, then stop, as another shock seemed to take the earth from under their feet, and many of them flung themselves face downward on the ground in an agony of fear.

Two or three minutes seemed to pass before the fugitives found their voices. Then the screams of women and the wild cries of men rent the air, and with one impulse the terror-stricken host fled toward the parks, to get themselves as far as possible from the tottering and falling walls. These speedily became packed with people, most of them in the night clothes in which they had leaped or been flung from their beds, screaming and moaning at the little shocks that at intervals followed the great one. The dawn was just breaking. The gas and electric mains were gone and the street lamps were all out. The sky was growing white in the east, but before the sun could fling his early rays from the horizon there came another light, a lurid and threatening one, that of the flames that had begun to rise in the warehouse district.

The braver men and those without families to watch over set out for this endangered region, half dressed as they were. In the early morning light they could see the business district below them, many of the buildings in ruins and the flames showing redly in five or six places. Through the streets came the fire engines, called from the outlying districts by a general alarm. The firemen were not aware as yet that no water was to be had.

THE PANIC IN THE SLUMS.

On Portsmouth Square the panic was indescribable. This old tree plaza, about which the early city was built, is now in the centre of Chinatown, of the Italian district and of the "Barbary Coast," the "Tenderloin" of the Western metropolis. It is the chief slum district of the city. The tremor here ran up the Chinatown hill and shook down part of the crazy buildings on its southern edge. It brought ruin also to some of the Italian tenements. Portsmouth Square became the refuge of the terrified inhabitants. Out from their underground burrows like so many rats fled the Chinese, trembling in terror into the square, and seeking by beating gongs and other noise-making instruments to scare off the underground demons. Into the square from the other side came the Italian refugees. The panic became a madness, knives were drawn in the insanity of the moment, and two Chinamen were taken to the morgue, stabbed to death for no other reason than pure madness. Here on one side dwelt 20,000 Chinese, and on the other thousands of Italians, Spaniards and Mexicans, while close at hand lived the riff-raff of the "Barbary Coast."

Seemingly the whole of these rushed for that one square of open ground, the two streams meeting in the centre of the square and heaping up on its edges. There they squabbled and fought in the madness of panic and despair, as so many mad wolves might have fought when caught in the red whirl of a prairie fire, until the soldiers broke in and at the bayonet's point brought some semblance of order out of the confusion of panic terror.

This scene in Portsmouth Square but illustrated the madness of fear everywhere prevailing. On every side thousands were fleeing from the roaring furnace that minute by minute seemed to extend its boundaries.

THE FLIGHT FOR SAFETY.

In the awful scramble for safety the half-crazed survivors disregarded everything but the thought of themselves and their property. In every excavation and hole throughout the north beach householders buried household effects, throwing them into ditches and covering the holes. Attempts were made to mark the graves of the property so that it could be recovered after the flames were appeased.

The streets were filled with struggling people, some crying and weeping and calling for missing loved ones. Crowding the sidewalks were thousands of householders attempting to drag some of their effects to places of safety. In some instances men with ropes were dragging trunks, tandem style, while others had sewing machines strapped to the trunks. Again, women were rushing for the hills, carrying on their arms only the family cat or a bird cage.

There were two ideas in the minds of the fugitives, and in many cases these two only. One of these was to escape to the open ground of Golden Gate Park and the Presidio reservation; the other was to reach the ferry and make their way out of the seemingly doomed city.

At the ferry building a crowd numbering thousands gathered, begging for food and transportation across the bay. Hundreds had not even the ten cents fare to Oakland. Most of the refugees at this point were Chinamen and Italians, who had fled from their burned tenements with little or no personal property.

Residents of the hillsides in the central portion of the city seemingly were safe from the inferno of flames that was consuming the business section. They watched the towering mounds of flames, and speculated as to the extent of the territory that was doomed. Suddenly there was whispered alarm up and down the long line of watchers, and they hurried away to drag clothing, cooking utensils and scant provisions through the streets. From Grant Avenue the procession moved westward. Men and women dragged trunks, packed huge bundles of blankets, boxes of provisions—everything. Wagons could not be hired except by paying the most extortionate rates.

"Thank Heaven for the open space of the Presidio and for Golden Gate Park!" was the unspoken thank-offering of many hearts. The great park, with its thousand and more acres of area, extending from the thinly populated part of the city across the sand dunes to the Pacific, seemed in that awful hour a God-given place of refuge. Near it and extending to the Golden Gate channel is the Presidio military reservation, containing 1,480 acres, and with only a few houses on its broad extent. Here also was a place of safety, provided that the forests which form a part of its area did not burn.

THE EXODUS FROM THE BURNING CITY.

To these open spaces, to the suburbs, in every available direction, the fugitives streamed, in thousands, in tens of thousands, finally in hundreds of thousands, safety from those towering flames, from the tottering walls of their dwellings, from a possible return of the earthquake, their one overmastering thought. There were many persons with scanty clothing, women in underskirts and thin waists and men in shirt sleeves. Many women carried children, while others wheeled baby carriages. It was a strange and weird procession, that kept up unceasingly all that dreadful day and through the night that followed, as the all-conquering flames spread the area of terror.

At intervals news came of what was doing behind the smoke cloud. The area of the flames spread all night. People who had decided that their houses were outside of the dangerous area and had decided to pass the night, even after the terrible experience of the shake-up, under their roofs, hourly gave up the idea and struggled to the parks. There they lay in blankets, their choicest valuables by their sides, and the soldiers kept watch and order. Many lay on the bare grass of the park, with nothing between them and the chill night air. Fortunately, the weather was clear and mild, but among those who lay under the open sky were men and women who were delicately reared, accustomed all their lives to luxurious surroundings, and these must have suffered severely during that night of terror.

The fire was going on in the district south of them, and at intervals all night exhausted fire-fighters made their way to the plaza and dropped, with the breath out of them, among the huddled people and the bundles of household goods. The soldiers, who were administering affairs with all the justice of judges and all the devotion of heroes, kept three or four buckets of water, even from the women, for these men, who continued to come all the night long. There was a little food, also kept by the soldiers for these emergencies, and the sergeant had in his charge one precious bottle of whisky, from which he doled out drinks to those who were utterly exhausted.

But there was no panic. The people were calm, stunned. They did not seem to realize the extent of the calamity. They heard that the city was being destroyed; they told each other in the most natural tone that their residences were destroyed by the flames, but there was no hysteria, no outcry, no criticism.

The trip to the hills and to the water front was one of terrible hardship. Famishing women and children and exhausted men were compelled to walk seven miles around the north shore in order to avoid the flames and reach the ferries. Many dropped to the street under the weight of their loads, and willing fathers and husbands, their strength almost gone, strove to pick up and urge them forward again.

In the panic many mad things were done. Even soldiers were obliged in many instances to prevent men and women, made insane from the misfortune that had engulfed them, from rushing into doomed buildings in the hope of saving valuables from the ruins. In nearly every instance such action resulted in death to those who tried it. At Larkin and Sutter Streets, two men and a woman broke from the police and rushed into a burning apartment house, never to reappear.

The rush to the parks and the dunes was followed in the days that followed by as wild a rush to the ferries, due to the mad desire to escape anywhere, in any way, from the burning city.

THE WILD RUSH TO THE FERRIES.

At the ferry station on Wednesday night there was much confusion. Mingled in an inextricable mass were people of every race and class on earth. A common misfortune and hunger obliterated all distinctions. Chinese, lying on pallets of rags, slept near exhausted white women with babies in their arms. Bedding, household furniture of every description, pet animals and trinkets, luggage and packages of every sort packed almost every foot of space near the ferry building. Men spread bedding on the pavement and calmly slept the sleep of exhaustion, while all around a bedlam of confusion reigned.

Many of those who sought the ferry on that fatal Wednesday met a solid wall of flames extending for squares in length and utterly impassable. In their half insane eagerness to escape some of them would have rushed into fatal danger but for the soldiers, who guarded the fire line and forced them back. Only those reached the ferry who had come in precedence of the flames, or who made a long detour to reach that avenue of flight. When the news came to the camps of refugees that it was safe to cross the burned area a procession began from the Golden Gate Park across the city and down Market Street, the thoroughfare which had long been the pride of the citizens, and a second from the Presidio, along the curving shore line of the north bay, thence southward along the water front. Throughout these routes, eight miles long, a continuous flow of humanity dragged its weary way all day and far into the night amidst hundreds of vehicles, from the clumsy garbage cart to the modern automobile. Almost every person and every vehicle carried luggage. Drivers of vehicles were disregardful of these exhausted, hungry refugees and drove straight through the crowd. So dazed and deadened to all feeling were some of them that they were bumped aside by carriage wheels or bumped out of the way by persons.

SCENES OF HUMOR AND PATHOS.

As already stated, the scene had its humorous as well as its pathetic side, and various amusing stories are told by those who were in a frame of mind to notice ludicrous incidents in the horrors of the situation. Two race track men met in the drive.

"Hello, Bill; where are you living now?" asked one.

"You see that tree over there—that big one?" said Bill. "Well, you climb that. My room is on the third branch to the left," and they went away laughing.

Another observer tells these incidents of the flight: "I saw one big fat man calmly walking up Market Street, carrying a huge bird cage, and the cage was empty. He seemed to enjoy looking at the wrecked buildings. Another man was leading a huge Newfoundland dog and carrying a kitten in his arms. He kept talking to the kitten. On Fell Street I noticed an old woman, half dressed, pushing a sewing machine up the hill. A drawer fell out, and she stopped to gather the fallen spools. Poor little seamstress, it was now her all."

A more amusing instance of the spirit of saving is that told by another narrator, who says that he saw a lone woman patiently pushing an upright piano along the pavement a few inches at a time. Evidently in this case, too, it was the poor soul's one great treasure on earth.

He also tells of a guest berating the proprietor of a hotel, a few minutes after the shock, because he had not obeyed orders to call him at five o'clock. He vowed he would never stop at that house again, a vow he might well keep, as the house is no more.

In one room where two girls were dressing the floor gave way and one of them disappeared.

"Where are you, Mary?" screamed her companion.

"Oh, I'm in the parlor," said Mary calmly, as she wriggled out of the mass of plaster and mortar below.

At the handsome residence of Rudolph Spreckels, the wealthy financier, the lawn was riven from end to end in great gashes, while the ornamental Italian rail leading to the imposing entrance was a battered heap. But the family, with a philosophy notable for the occasion, calmly set up housekeeping on the sidewalk, the women seated in armchairs taken from the mansion and wrapped in rugs and coverlets, the silver breakfast service was laid out on the stone coping and their morning meal spread out on the sidewalk. This scene was repeated at other houses of the wealthy, the families too fearful of another shock to venture within doors.

Another story of much interest in this connection is told. On Friday afternoon, two days and some hours after the scene just narrated, Mrs. Rudolph Spreckels presented her husband with an heir on the lawn in front of their mansion, while the family were awaiting the coming of the dynamite squad to blow up their magnificent residence. An Irish woman who had been called in to play the part of midwife at a birth elsewhere on Saturday, made a pertinent comment after the wee one's eyes were opened to the walls of its tent home.

"God sends earthquakes and babies," she said, "but He might, in His mercy, cut out sending them both together."

There were many pathetic incidents. Families had been sadly separated in the confusion of the flight. Husbands had lost their wives—wives had lost their husbands, and anxious mothers sought some word of their children—the stories were very much the same. One pretty looking woman in an expensive tailor-made costume badly torn, had lost her little girl.

"I don't think anything has happened to her," said she, hopefully. "She is almost eleven years old, and some one will be sure to take her in and care for her; I only want to know where she is. That is all I care about now."

A well-known young lady of good social position, when asked where she had spent the night, replied: "On a grave."

"I thank God, I thank Uncle Sam and the people of this nation," said a woman, clad in a red woolen wrapper, seated in front of a tent at the Presidio nursing one child and feeding three others from a board propped on two bricks. "We have lost our home and all we had, but we have never been hungry nor without shelter."

The spirit of '49 was vital in many of the refugees. One man wanted to know whether the fire had reached his home. He was informed that there was not a house standing in that section of the city. He shrugged his shoulders and whistled.

"There's lots of others in the same boat," as he turned away.

"Going to build?" repeated one man, who had lost family and home inside of two hours. "Of course, I am. They tell me that the money in the banks is still all right, and I have some insurance. Fifteen years ago I began with these," showing his hands, "and I guess I'm game to do it over again. Build again, well I wonder."

Among the many pathetic incidents of the disaster was that of a woman who sat at the foot of Van Ness Avenue on the hot sands on the hillside overlooking the bay east of Fort Mason, with four little children, the youngest a girl of three, the eldest a boy of ten years. They were destitute of water, food and money.

The woman had fled, with her children, from a home in flames in the Mission Street district, and tramped to the bay in the hope of sighting the ship which she said was about due, of which her husband was the captain.

"He would know me anywhere," she said. And she would not move, although a young fellow gallantly offered his tent, back on a vacant lot, in which to shelter her children.

THE GOLDEN GATE CAMP.

In the Golden Gate Park there was the most woefully grotesque camp of sufferers imaginable. There was no caste, no distinction of rich and poor, social lines had been obliterated by the common misfortune, and the late owners of property and wealth were glad to camp by the side of the day laborer. As for shelter, there were a few army tents and some others which afforded a fair degree of comfort, but nine out of ten are the poorest suggestions of tents made out of bedclothes, rugs, raincoats and in some cases of lace curtains. None of the tents or huts has a floor, and it is impossible to see how a large number of women and children can escape the most disastrous physical effects.

The unspeakable chaos that prevailed was apparent in no way more than in the system, or lack of system, of registration and location. At the entrance to Golden Gate Park stands a billboard, twenty feet high and a hundred feet long. Originally it bore the praises of somebody's beer. Covering this billboard, to a height of ten or twelve feet, were slips of paper, business cards, letter heads and other notices, addressed to "Those interested," "Friends and relatives," or to some individual, telling of the whereabouts of refugees.

One notice read: "Mrs. Rogers will find her husband in Isidora Park, Oakland. W. H. Rogers." Another style was this: "Sue, Harry and Will Sollenberger all safe. Call at No. 250 Twenty-seventh Avenue."

There were thousands of these dramatic notices on this billboard, and one larger than the others read: "Death notices can be left here; get as many as possible."

Another method of finding friends and relatives was by printing notices on vehicles. On the side curtains of a buggy being driven to Golden Gate Park was the following sign: "I am looking for I. E. Hall."

That searchers for lost ones might have the least trouble, all the tents, here known as camps, were tagged with the names or numbers. For instance, one tent of bed quilts carried this sign: "No. 40 Bush Street camp."

Most of the tents were merely named for the family name of the occupants, the former streets number usually being given. But these tent tags told a wonderful story of human nature. A small army tent bore the name, "Camp Thankful," the one next to it was placarded "Camp Glory" and a few feet farther on an Irishman had posted the sign "Camp Hell."

The cooking was all done on a dozen bricks for a stove, with such utensils as may usually be picked up in the ordinary residential alley. But in all of the camps the badge of the eternal feminine was to be found in the form of small pieces of broken mirrors, or hand mirrors fastened to trees or tent walls, in some cases the polished bottom of a tomato can serving the purposes of the feminine toilet.

One woman, in whose improvised tent screeched a parrot, sat ministering to the wounds of the other family pet, a badly singed cat. The number of canaries, parrots, dogs and cats was one of the amusing features of the disaster.

Among the interesting and thrilling incidents of the disaster is that connected with the telegraph service. For many hours virtually all the news from San Francisco came over the wires of the Postal Telegraph Company. The Postal has about fifteen wires running into San Francisco. They go under the bay in cables from Oakland, and thence run underground for several blocks down Market Street to the Postal building. About forty operators are employed to handle the business, but evidently there was only about one on duty when the earthquake began.

What became of him nobody knows. But he seems to have

sent the first word of the disaster. It came over the Postal wires about nine o'clock, just when the day's business had started in the East. It will long be preserved in the records of the company. This was the dispatch:

"There was an earthquake hit us at 5.13 this morning, wrecking several buildings and wrecking our offices. They are carting dead from the fallen buildings. Fire all over town. There is no water and we lost our power. I'm going to get out of office, as we have had a little shake every few minutes, and it's me for the simple life."

"R., San Francisco, 5.50 A. M."

"Mr. R." evidently got out, for there was nothing doing for a brief interval after that. The operator in the East pounded and pounded at his key, but San Francisco was silent. The Postal people were wondering if it was all the dream of some crazy operator or a calamity, when the wire woke up again. It was the superintendent of the San Francisco force this time.

"We're on the job, and are going to try and stick," was the way the first message came from him.

This was what came over the wire a little later:

"Terrific earthquake occurred here at 5.13 this morning. A number of people were killed in the city. None of the Postal people were killed. They are now carting the dead from the fallen buildings. There are many fires, with no one to fight them. Postal building roof wrecked, but not entire building."

The fire got nearer and nearer to the Postal building. All of the water mains had been destroyed around the building, the operators said, and there was no hope if the fire came on. They also said that they could hear the sound of dynamite blowing up buildings. All this time the operators were sticking to their posts and sending and receiving all the business the wires could stand. At 12.45 the wire began to click again with a message for the little group of waiting officials.

This message came in jerks: "Fire still coming up Market Street. It's one block from the Post Office now; back of the Palace Hotel is a furnace. I am afraid that the Grand Hotel and the Palace Hotel will get it soon. The Southern Pacific offices on California Street are safe, so far, but can't tell what will happen. California Street is on fire. Almost everything east of Montgomery Street and north of Market Street is on fire now."

There was a pause, then: "We are beginning to pack up our instruments."

"Instruments are all packed up, and we are ready to run," was another message. It was evident that just one instrument had been left connected with the world outside. In about ten minutes it began to click. Those who knew the telegraphers' language caught the word "Good-bye," and then the ticks stopped.

At the end of an hour the instrument in the office began to click again. It was from an electrician by the name of Swain.

"I'm back in the building, but they are dynamiting the building next door, and I've got to get out," was the way his message was translated. Dynamite ended the story, and the Postal's domicile in San Francisco ceased to exist.

CHAPTER VI.

Facing Famine and Praying for Relief.

FRIGHTFUL was the emergency of the vast host of fugitives who fled in terror from the blazing city of San Francisco to the open gates of Golden Gate Park and the military reservation of the Presidio. Food was wanting, scarcely any water was to be had, death by hunger and thirst threatened more than a quarter million of souls thus driven without warning from their comfortable and happy homes and left without food or shelter. Provisions, shelter tents, means of relief of various kinds were being hurried forward in all haste, but for several days the host of fugitives had no beds but the bare ground, no shelter but the open heavens, scarcely a crumb of bread to eat, scarcely a gill of water to drink. Those first days that followed the disaster were days of horror and dread. Rich and poor were mingled together, the delicately reared with the rough sons of toil to whom privation was no new experience.

Those who had food to sell sought to take advantage of the necessities of the suffering by charging famine prices for their supplies, but the soldiers put a quick stop to this. When Thursday morning broke, lines of buyers formed before the stores whose supplies had not been commandeered. In one of these, the first man was charged 75 cents for a loaf of bread. The corporal in charge at that point brought his gun down with a slam.

"Bread is 10 cents a loaf in this shop," he said.

It went. The soldier fixed the schedule of prices a little higher than in ordinary times, and to make up for that he forced the store-keeper to give free food to several hungry people in line who had no money to pay. In several other places the soldiers used the same brand of horse sense.

A man with a loaf of bread in his hand ran up to a policeman on Washington Street. "Here," he said, "this man is trying to charge me a dollar for this loaf of bread. Is that fair?"

"Give it to me," said the policeman. He broke off one end of it and stuck it in his mouth. "I am hungry myself," he said when he had his mouth clear. "Take the rest of it. It's appropriated."

As an example of the prices charged for food and service by the unscrupulous, we may quote the experience of a Los Angeles millionaire named John Singleton, who had been staying a day or two at the Palace Hotel. On Wednesday he had to pay $25 for an express wagon to carry himself, his wife and her sister to the Casino, near Golden Gate Park, and on Thursday was charged a dollar apiece for eggs and a dollar for a loaf of bread. Others tell of having to pay $50 for a ride to the ferry.

One of the refugees on the shores of Lake Herced Thursday morning spied a flock of ducks and swans which the city maintained there for the decoration of the lake. He plunged into the lake, swam out to them and captured a fat drake. Other men and boys saw the point and followed. The municipal ducks were all cooking in five minutes.

The soldiers were prompt to take charge of the famine situation, acting on their own responsibility in clearing out the supplies of the little grocery stores left standing and distributing them

among the people in need. The principal food of those who remained in the city was composed of canned goods and crackers.
The refugees who succeeded in getting out of San Francisco were
met as soon as they entered the neighboring towns by representa-
tives of bakers who had made large supplies of bread, and who
immediately dealt them out to the hungry people.

THE FOOD QUESTION URGENT.

But the needs of the three hundred thousand homeless and
hungry people in the city could not be met in this way, and immediate supplies in large quantities were necessary to prevent a reign
of famine from succeeding the ravages of the fire. Danger from
thirst was still more insistent than that from hunger. There was
some food to be had, bakeries were quickly built within the military
reservation there, and General Funston announced that rations
would soon reach the city and the people would be supplied from
the Presidio. But there was scarcely any water to relieve the
thirst of the suffering. Water became the incessant cry of firemen
and people alike, the one wanting it to fight the fire, the other to
drink, but even for the latter the supply was very scant. There
was water in plenty in the reservoirs, but they were distant and
difficult to reach, and all night of the day succeeding the earth
shock wagons mounted with barrels and guarded by soldiers drove
through the park doling out water. There was a steady crush
around these wagons, but only one drink was allowed to a person.

Toward midnight a black, staggering body of men began to
weave through the entrance. They were volunteer fire-fighters,
looking for a place to throw themselves down and sleep. These

men dropped out all along the line, and were rolled out of the drive-
ways by the troops.   There was much splendid unselfishness here.
Women gave up their blankets and sat up or walked about all night
to cover the exhausted men who had fought fire until there was no
more fight in them.

The common destitution and suffering had, as we have said,
wiped out all social, financial and racial distinctions.   The man who
last Tuesday was a prosperous merchant was obliged to occupy
with his family a little plot of ground that adjoined the open-air
home of a laborer.   The white man of California forgot his antipathy
to the Asiatic race, and maintained friendly relations with his new
Chinese and· Japanese neighbors.   The society belle who Tuesday
night was a butterfly of fashion at the grand opera performance
now assisted some factory girl in the preparation of humble daily
meals.   Money had little value.   The family that had had foresight
to lay in the largest stock of foodstuffs on the first day of disaster
was rated highest in the scale of wealth.

A few of the families that could secure wagons were possess-
ors of cook stoves, but over 95 per cent. of the refugees did their
cooking on little campfires made of brick or stone.   Battered kitchen
utensils that the week before would have been regarded as useless
had become articles of high value.   In fact, man had come back to
nature and all lines of caste had been obliterated, while the very
thought of luxury had disappeared.   It was, in the exigency of the
moment, considered good fortune to have a scant supply of the
barest necessaries of life.

As for clothing, it was in many cases of the scantiest, while
numbers of the people had brought comfortable clothing and bedding.   Many others had fled in their night garbs, and comparatively

few of these had had the self-possession to return and don their daytime clothes. As a result there had been much improvisation of garments suitable for life in the open air, and as the days went on many of the women arrayed themselves in home-made bloomer costumes, a sensible innovation under the circumstances and in view of the active outdoor work they were obliged to perform.

The grave question to be faced at this early stage was: How soon would an adequate supply of food arrive from outside points to avert famine? Little remained in San Francisco beyond the area swept by the fire, and the available supply could not last more than a few days. Fresh meat disappeared early on Wednesday and only canned foods and breadstuffs were left. All the foodstuffs coming in on the cars were at once seized by order of the Mayor and added to the scanty supply, the names of the consignees being taken that this material might eventually be paid for. The bakers agreed to work their plants to their utmost capacity and to send all their surplus output to the relief committee. By working night and day thousands of loaves could be provided daily. A big bakery in the saved district started its ovens and arranged to bake 50,000 loaves before night. The provisions were taken charge of by a committee and sent to the various depots from which the people were being fed. Instructions were issued by Mayor Schmitz on Thursday to break open every store containing provisions and to distribute them to the thousands under police supervision. A policeman reported that two grocery stores in the neighborhood were closed, although the clerks were present. "Smash the stores open," ordered the Mayor, "'and guard them." In towns across the bay the master bakers have met and fixed the price of bread at 5 cents the loaf, with the understanding that they will refuse to sell to retailers who attempt to charge famine prices. The committee of citizens in charge of the situation in the stricken city proposed to use every effort to keep food down to the ordinary price and check the efforts of speculators, who in one instance charged as much as $3.50 for two loaves of bread and a can of sardines. Orders were issued by the War Department to army officers to purchase at Los Angeles immediately 200,000 rations and at Seattle 300,000 rations and hurry them to San Francisco. The department was informed that there were 120,000 rations at the Presidio, that thousands of refugees were being sheltered there and that the army was feeding them. One million rations already had been started to San Francisco by the department. But in view of the fact that there were 300,000 fugitives to be fed the supply available was likely to be soon exhausted.

FOOD FOR THE HUNGRY.

Such was the state of affairs at the end of the second day of the great disaster. But meanwhile the entire country had been aroused by the tidings of the awful calamity, the sympathetic instinct of Americans everywhere was awakened, and it was quickly made evident that the people of the stricken city would not be allowed to suffer for the necessaries of life. On all sides money was contributed in large sums, the United States Government setting the example by an immediate appropriation of $1,000,000, and in the briefest possible interval relief trains were speeding toward the stricken city from all quarters, carrying supplies of food, shelter tents and other necessaries of a kind that could not await deliberate action.

Shelter was needed almost as badly as food, for a host of the refugees had nothing but their thin clothing to cover them, and, though the weather at first was fine and mild, a storm might come at any time. In fact, a rain did come, a severe one, early in the week after the disaster, pouring nearly all night long on the shivering campers in the parks, wetting them to the skin and soaking through the rudely improvised shelters which many of the refugees had put up. A few days afterward came a second shower, rendering still more evident the need of haste in providing suitable shelter.

All this was foreseen by those in charge, and the most strenuous efforts were made to provide the absolute necessities of life. Huge quantities of supplies were poured into the city. From all parts of California trainloads of food were rushed there in all haste. A steamer from the Orient laden with food reached the city in its hour of need; another was dispatched in all haste from Tacoma bearing $25,000 worth of food and medical supplies, ordered by Mayor Weaver, of Philadelphia, as a first installment of that city's contribution. Money was telegraphed from all quarters to the Governor of California, to be expended for food and other supplies, and so prompt was the response to the insistent demand that by Saturday all danger of famine was at an end; the people were being fed.

WATER FOR THE THIRSTY.

The broken waterpipes were also repaired with all possible haste, the Spring Valley Water Company putting about one thousand men at work upon their shattered mains, and in a very brief

EUGENE E. SCHMITZ, MAYOR OF SAN FRANCISCO.
Mayor Schmitz, by his untiring efforts and great executive ability, brought order out of chaos.

CHASMS IN A STREET OF SAN FRANCISCO.
The thick asphalt pavement was opened in deep fissures by the first earthquake. At the lower left hand appears one of the broken water mains, which cut off all water supply and permitted unchecked play to the flames.

THE RUINED PALACE HOTEL.

This photograph was taken on Market Street, and shows in the foreground the magnificent Palace Hotel, where many of the guests barely escaped from the building with their lives. The Monadnock Building appears on its right, and next to it the tall pile of the "Call" Building.

A PANORAMA OF DEVASTATION. This photograph was taken from Nob Hill, looking toward the Bay and Goat Island on one of the principal thoroughfares, and shows but few buildings left more than one story high above the ground in the center of the principal business district.

53

time water began to flow freely in many parts of the residence section and the great difficulty of obtaining food and water was practically at an end. Never in the history of the country has there been a more rapid and complete demonstration of the resourcefulness of Americans than in the way this frightful disaster was met.

Food, water and shelter were not the only urgent needs. At first there was absolutely no sanitary provision, and the danger of an epidemic was great. This was a peril which the Board of Health addressed itself vigorously to meet, and steps for improving the sanitary conditions were hastily taken. Quick provision for sheltering the unfortunates was also made. Eight temporary structures, 150 feet in length by 28 feet wide and 13 feet high, were erected in Golden Gate Park, and in these sheds thousands found reasonably comfortable quarters. This was but a beginning. More of these buildings were rapidly erected, and by their aid the question of shelter was in part solved. The buildings were divided into compartments large enough to house a family, each compartment having an entrance from the outside. This work was done under the control of the engineering department of the United States army, which had taken steps to obtain a full supply of lumber and had put 135 carpenters to work. Those of the refugees who were without tents were the first to be provided for in these temporary buildings.

THE CAMPS IN THE PARKS.

To those who made an inspection of the situation a few days after the earthquake, the hills and beaches of San Francisco looked like an immense tented city. For miles through the park and along

the beaches from Ingleside to the sea wall at North Beach the homeless were camped in tents—makeshifts rigged up from a few sticks
of wood and a blanket or sheet. Some few of the more fortunate
secured vehicles on which they loaded regulation tents and were,
therefore, more comfortably housed than the great majority. Golden
Gate Park and the Panhandle looked like one vast campaign ground.
It is said that fully 100,000 persons, rich and poor alike, sought
refuge in Golden Gate Park alone, and 200,000 more homeless ones
located at the other places of refuge.

At the Presidio military reservation, where probably 50,000
persons were camped, affairs were conducted with military precision.
Water was plentiful and rations were dealt out all day long. The
refugees stood patiently in line and there was not a murmur. This
characteristic was observable all over the city. The people were
brave and patient, and the wonderful order preserved by them proved
of great assistance. In Golden Gate Park a huge supply station
had been established and provisions were dealt out.

Six hundred men from the Ocean Shore Railway arrived on
Saturday night with wagons and implements to work on the sewer
system. Inspectors were kept going from house to house, examining
chimneys and issuing permits to build fires. In fact, activity manifested itself in all quarters in the attempt to bring order out of
confusion, and in an astonishingly short time the tented city was
converted from a scene of wretched disorder into one of order and
system.

At Jefferson Park were camped thousands of people of every
class in life. On the western edge of this park is the old Scott
house, where Mrs. McKinley lay sick for two weeks in 1901. Three
times a day the people all gathered in line before the provision

wagons for their little handouts. "Yesterday," says an observer, "I saw, in order before the wagons, a Lascar sailor in his turban, about as low a Chinatown bum as I ever set eyes on, a woman of refined appearance, a barefooted child, two Chinamen, and a pretty girl. They were squeezed up together by the line, which extended for a quarter of a mile. It is civilization in the bare bones.

"The great and rich are on a level with the poor in the struggle for bare existence, and over them all is the perfect, unbroken discipline of the soldiery. They came into the city and took charge on an hour's notice, they saved the city from itself in the three days of hell, and but for them the city, even with enough provisions to feed them in the stores and warehouses, must have gone hungry for lack of distributive organization."

COMEDY AND PATHOS IN THE BREAD LINE.

At one of the parks on Tuesday morning a handsomely dressed woman with two children at her skirts stood in a line of many hundreds where supplies were being given out. She took some uncooked bacon, and as she reached for it jewels sparkled on her fingers. One of the tots took a can of condensed milk, the other a bag of cakes.

"I have money," she said, "'if I could get it and use it. I have property, if I could realize on it. I have friends, if I could get to them. Meantime I am going to cook this piece of bacon on bricks and be happy."

She was only one of thousands like her.

In a walk through the city this note of cheerfulness of the people in the face of an almost incredible week of horror was to a correspondent the mitigating element to the awfulness of disaster.

In the streets of the residential district in the western addition, which the fire did not reach, women of the houses were cooking meals on the pavement. In most cases they had moved out the family ranges, and were preparing the food which they had secured from the Relief Committee.

Out on Broderick street, near the Panhandle, a piano sounded. It was nigh ten o'clock and the stars were shining after the rain. Fires gleamed up and down through the shrubbery and the refugees sat huddled together about the flames, with their blankets about their heads, Apache-like, in an effort to dry out after the wetting of the afternoon. The piano, dripping with moisture, stood on the curb, near the front of a cottage which had been wrecked by the earthquake.

A youth with a shock of red hair sat on a cracker box and pecked at the ivories. "Home Ain't Nothing Like This" was thrummed from the rusting wires with true vaudeville dash and syncopation. "Bill Bailey," "Good Old Summer Time," "Dixie" and "In Toyland" followed. Three young men with handkerchiefs wrapped about their throats in lieu of collars stood near the pianist and with him lifted up their voices in melody. The harmony was execrable, the time without excuse, but the songs ran through the trees of the Panhandle, and the crows, forgetting their misery for a time, joined the strange chorus.

The people had their tales of comedy, one being that on the morning of the fire a richly dressed woman who lived in one of the aristocratic Sutter Street apartments came hurrying down the street, faultlessly gowned as to silks and sables, save that one dainty foot was shod with a high-heeled French slipper and the other was incased in a laborer's brogan. They say that as she walked she careened like a bark-rigged ship before a typhoon.

An hour spent behind the counter of the food supply depot in the park tennis court yielded rich reward to the seeker after the outlandish. The tennis court was piled high with the plunder of several grocery stores and the cargoes of many relief cars. A square cut in the wire screen permitted of the insertion of a counter, behind which stood members of the militia acting as food dispensers. Before the improvised window passed the line of refugees, a line which stretched back fully 300 yards to Speedway track.

"I want a can of condensed cream, so I can feed my baby and my dog," said a large, florid-faced woman in a gaudy kimono, "and I don't care for crackers, but you can throw in some potted chicken if you have it."

"What's in that bottle over there?" queried the next applicant. "Tomato ketchup? Well, of all the luck! Say, young man, just give me three."

A little gray-haired woman in an India shawl peered timorously through the window. "Just a little bit of anything you may have handy, please," she whispered, and she cast a careful eye about to see of any of her neighbors had recognized her standing there in the "bread line."

"Yesterday, at the Western Union office," says one writer, "I saw a woman drive up in a large motor car and beg that the telegram on which a boy had asked a delivery fee of twenty-five cents be handed to her. She said she had not a penny and did not know when she would have any money, but that as soon as she had any she would pay for the message. It was given to her, and the manager told me that there were hundreds of similar cases."

Many weddings resulted from the disaster. Women driven out of their homes and left destitute, appealed to the men to whom they were engaged, and immediate marriages took place. After the first day of the disaster an increase in the marriage licenses issued was noticed by County Clerk Cook. This increase grew until seven marriage licenses were issued in an hour.

"I don't live anywhere," was the answer given in many cases when the applicant for a license was asked the locality of his residence. "I used to live in San Francisco."

Births seem to have been about as common as marriages, in one night five children being born in Golden Gate Park. In Buena Vista Park eight births were recorded and others elsewhere, the population being thus increased at a rate hardly in accordance with the exigencies of the situation.

THE EXODUS FROM SAN FRANCISCO.

We have spoken only of the camps of refugees within the municipal limits of San Francisco. But in addition to these was the multitude of fugitives who made all haste to escape from that city. This was with the full consent of the authorities, who felt that every one gone lessened the immediate weight upon themselves, and who issued a strict edict that those who went must stay, that there could be no return until a counter edict should be made public.

From the start this was one of the features of the situation. Down Market Street, once San Francisco's pride, now leading through piles of tottering walls, piles of still hot bricks and twisted iron and heaps of smouldering debris, poured a huge stream of pedestrians. Men bending under the weight of great bundles pushed baby carriages loaded with bric-a-brac and children. Women toiled along with their arms full, but a large proportion were able to ride, for the relief corps had been thoroughly organized and wagons were being pressed into service from all sides.

In constant procession they moved toward the ferry, whence the Southern Pacific was transporting them with baggage free wherever they wished to go. Automobiles meanwhile shot in all directions, carrying the Red Cross flag and usually with a soldier carrying a rifle in the front seat. They had the right of way every-where, carrying messages and transporting the ill to temporary hospitals and bearing succor to those in distress.

Oakland, the nearest place of resort, on the bay shore opposite San Francisco, soon became a great city of refuge, fugitives gath-ering there until 50,000 or more were sheltered within its charitable limits. Having suffered very slightly from the earthquake that had wrecked the great city across the bay, it was in condition to offer shelter to the unfortunate. All day Wednesday and Thursday a stream of humanity poured from the ferries, every one carrying personal baggage and articles saved from the conflagration. Hun-dreds of Chinese men, women and children, all carrying baggage to the limit of their strength, made their way into the limited Chinatown of Oakland.

Multitudes of persons besieged the telegraph offices, and the crush became so great that soldiers were stationed at the doors to keep them in line and allow as many as possible to find standing room at the counters. Messages were stacked yards high in the offices waiting to be sent throughout the world. Every boat from San Francisco brought hundreds of refugees, carrying luggage and bedding in large quantities. Many women were bareheaded and all showed fatigue as the result of sleeplessness and exposure to the chill air. Hundreds of these persons lined the streets of Oakland, waiting for some one to provide them with shelter, for which the utmost possible provision was quickly made. No one was allowed to go hungry in Oakland and few lacked shelter. At the Oakland First Presbyterian Church 1,800 were fed and 1,000 people were provided with sleeping accommodations. Pews were turned into beds. Cots stood in the aisles, in the gallery and in the Sunday school room. Every available inch of space was occupied by some substitute for a bed.

As the days wore on the number of refugees somewhat decreased. Although they still came in large numbers, many left on every train for different points. Requests for free transportation were investigated as closely as possible and all the deserving were sent away. Women and children and married men who wished to join their families in different parts of the State were given preference. The transportation bureau was on a street corner, where a man stood on a box and called the names of those entitled to passes.

Along the principal streets of Oakland there was a picturesque pilgrimage of former householders, who dragged or carried the meagre effects they had been able to save. The refugees who could not be cared for in Oakland made an exodus to Berkeley and other surrounding cities, where relief committees were actively at work. Utter despair was pictured on many faces, which showed the effects of sleepless days and nights, and the want of proper food.

Oakland was only one of the outside camps of refuge. At Berkeley over 6,000 refugees sought quarters, the big gymnasium of the State University being turned into a lodging house, while hundreds were provided with blankets to sleep in the open air under the University oaks. The students and professors of the University did all they could for their relief, and the Citizens' Relief Committee supplied them with food.

The same benevolent sympathy was manifested at all the places near the ruined city which had escaped disaster, this aid materially reducing that needed within San Francisco itself.

WORSHIP IN THE OPEN AIR.

Sunday dawned in San Francisco; Sunday in the camp of the refugees. On a green knoll in Golden Gate Park, between the conservatory and the tennis courts, a white-haired minister of the Gospel gathered his flock. It was the Sabbath day and in the turmoil and confusion the minister did not forget his duty. Two upright stakes and a cross-piece gave him a rude pulpit, and beside him stood a young man with a battered brass cornet. Far over the park stole a melody that drew hundreds of men and women from their tents. Of all denominations and all creeds, they gathered on that green knoll, and the men uncovered while the solemn voice repeated the words of a grand old hymn, known wherever men and women meet to worship the Lord:

"Other refuge have I none, hangs my helpless soul on Thee;
Leave, oh, leave me not alone, still support and comfort me!"

A moment before there had been shouting and confusion in the driveway where some red-striped artillerymen were herding a squad of gesticulating Chinamen as men herd sheep. The shouting died away as the minister's voice rose and fell and out of the stillness came the sobs of women. One little woman in blue was making no sound, but the tears were streaming down her cheeks. Her husband, a sturdy young fellow in his shirt sleeves, put his arm about her shoulders and tried to comfort her as the reading went on.

"All my trust on Thee is stayed; all my help from Thee I bring;
Cover my defenseless head with the shadow of Thy wing."

Then the cornet took up the air again and those helpless persons followed it in quivering tones, the white-haired man of God leading them with closed eyes. When the last verse was over, the minister raised his hands.

"Let us pray," said he, and his congregation sank down in the grass before him. It was a simple prayer, such a prayer as might be offered by a man without a home or a shelter over his head—and nothing left to him but an unshaken faith in his Creator.

"Oh, Lord, Thy ways are past finding out, but we still have faith in Thee. We know not why Thou hast visited these people and left them homeless. Thou knowest the reason of this desolation and of our utter helplessness. We call on Thee for help in the hour of our great need. Bless the people of this city, the sorrowing ones, the bereaved, gather them under Thy mighty wing and soothe aching hearts this day."

The women were crying again, and one big man dug his knuckles into his eyes without shame. The man who could have listened to such a prayer unmoved was not in Golden Gate Park that day.

CHAPTER VII.

The Frightful Loss of Life and Wealth.

WHILE multitudes escaped from toppling buildings and crashing walls in the dread disaster of that fatal Wednesday morning of April 18th in San Francisco, hundreds of the less fortunate met their death in the ruins, and horrifying scenes were witnessed by the survivors. Many of those who escaped had tales of terror to tell. Mr. J. P. Anthony, as he fled from the Ramona Hotel, saw a score or more of people crushed to death, and as he walked the streets at a later hour saw bodies of the dead being carried in garbage wagons and all kinds of vehicles to the improvised morgues, while hospitals and storerooms were already filled with the injured. Mr. G. A. Raymond, of Tomales, Cal., gives evidence to the same effect. As he rushed into the street, he says that the air was filled with falling stones and people around him were crushed to death on all sides.

Others gave testimony to the same effect. Samuel Wolf, of Salt Lake City, tells us that he saved one woman from death in the hotel. She was rushing blindly toward an open window, from which she would have fallen fifty feet to the stone pavement below. "On my way down Market Street," he says, "the whole side of a building fell out and came so near me that I was covered and blinded by the dust. Then I saw the first dead come by. They were piled up in an automobile like carcasses in a butcher's wagon, all bloody, with crushed skulls, broken limbs and bloody faces."

These are frightful stories, exaggerated probably from the nervous excitement of those terrible moments, as are also the following statements, which form part of the early accounts of the disaster. Thus we are told that "from a three-story lodging house at Fifth and Minna Streets, which collapsed Wednesday morning, more than seventy-five bodies were taken to-day. There are fifty other bodies in sight in the ruins. This building was one of the first to take fire on Fifth Street. At least 100 persons are said to have been killed in the Cosmopolitan, on Fourth Street. More than 150 persons are reported dead in the Brunswick Hotel, at Seventh and Mission Streets."

Another statement is to the effect that "at Seventh and Howard Streets a great lodging house took fire after the first shock, before the guests had escaped. There were few exits and nearly all the lodgers perished. Mrs. J. J. Munson, one of those in the building, leaped with her child in her arms from the second floor to the pavement below and escaped unhurt. She says she was the only one who escaped from the house. Such horrors as this were repeated at many points. B. Baker was killed while trying to get a body from the ruins. Other rescuers heard the pitiful wail of a little child, but were unable to get near the point from which the cry issued. Soon the onrushing fire ended the cry and the men turned to other tasks."

ESTIMATES OF THE DEATH LIST.

The questionable point in those statements is that the numbers of dead spoken of in these few instances exceed the whole number given in the official records issued two weeks after the disaster. Yet they go to illustrate the actual horrors of the case, and are of importance for this reason. As regards the whole number killed, in fact, there is not, and probably never will be, a full and accurate statement. While about 350 bodies had been recovered at the end of the second week, it was impossible to estimate how many lay buried under the ruins, to be discovered only as the work of excavation went on, and how many more had been utterly consumed by the flames, leaving no trace of their existence. The estimates of the probable loss of life ran up to 1,500 and more, while the injured were very numerous.

The shock of the earthquake, the pulse of deep horror to which it gave rise, the first wild impulse to flee for life, gave way in the minds of many to a feeling of intense sympathy as agonized cries came from those pinned down to the ruins of buildings or felled by falling bricks or stones, and as the sight of dead bodies incrimsoned with blood met the eyes of the survivors in the streets. From wandering aimlessly about, many of these went earnestly to work to rescue the wounded and recover the bodies of the slain. In this merciful work the police and the soldiers lent their aid, and soon there was a large corps of rescuers actively engaged.

BURYING THE DEAD.

Soon numbers were taken, alive or dead, from the ruins, passing vehicles were pressed into the service, and the labor of mercy went on rapidly, several buildings being quickly converted into temporary hospitals, while the dead were conveyed to the Mechanics' Pavilion and other available places. Portsmouth Square became for a time a public morgue. Between twenty and thirty corpses were laid side by side upon the trodden grass in the absence of more suitable accommodations. It is said that when the flames threatened to reach the square, the dead, mostly unknown, were removed to Columbia Square, where they were buried when danger threatened that quarter. Others were taken to the Presidio, and here the soldiers pressed into service all men who came near and forced them to labor at burying the dead, a temporary cemetery being opened there. So thick were the corpses piled up that they were becoming a menace, and early in the day the order was issued to bury them at any cost. The soldiers were needed for other work, so, at the point of rifles, the citizens were compelled to take to the work of burying. Some objected at first, but the troops stood no trifling, and every man who came within reach was forced to work. Rich men, unused to physical exertion, labored by the side of the workingmen digging trenches in which to bury the dead. The able-bodied being engaged in fighting the flames, General Funston ordered that the old men and the weaklings should take the work in hand. They did it willingly enough, but had they refused the troops on guard would have forced them. It was ruled that every man physically capable of handling a spade or a pick should dig for an hour. When the first shallow graves were ready the men, under the direction of the troops, lowered the bodies, several in a grave, and a strange burial began. The women gathered about crying. Many of them knelt while a Catholic priest read the burial service and pronounced absolution. All Thursday afternoon this went on.

In this connection the following stories are told:

Dr. George V. Schramm, a young medical graduate, said:

"As I was passing down Market Street with a new-found friend, an automobile came rushing along with two soldiers in it.

My doctor's badge protected me, but the soldiers invited my companion, a husky six-footer, to get into the automobile. He said:

" 'I don't want to ride, and have plenty of business to attend to.'

"Once more they invited him, and he refused. One of the soldiers pointed a gun at him and said:

" 'We need such men as you to save women and children and to help fight the fire.'

"The man was on his way to find his sister, but he yielded to the inevitable. He worked all day with the soldiers, and when released to get lunch he felt that he could conscientiously desert to go and find his own loved ones."

"Half a block down the street the soldiers were stopping all pedestrians without the official pass which showed that they were on relief business, and putting them to work heaving bricks off the pavement. Two dapper men with canes, the only clean people I saw, were caught at the corner by a sergeant, who showed great joy as he said:

" 'I give you time to git off those kid gloves, and then hustle, damn you, hustle!' The soldiers took delight in picking out the best dressed men and keeping them at the brick piles for long terms. I passed them in the shelter of a provision wagon, afraid that even my pass would not save me. Two men are reported shot because they refused to turn in and help."

Many of the dead, of course, will never be identified, though the names were taken of all who were known and descriptions written of the others. A story comes to us of one young girl who had followed for two days the body of her father, her only relative. It had been taken from a house on Mission Street to an undertaker's shop just after the quake. The fire drove her out with her charge, and it was placed in Mechanics' Pavilion. That went, and the body rested for a day at the Presidio, waiting burial. With many others, she wept on the border of the burned area, while the women cared for her.

VICTIMS TAKEN FROM THE RUINS.

On Friday eleven postal clerks, all alive, were taken from the debris of the Post Office. All at first were thought to be dead, but it was found that, although they were buried under the stone and timber, every one was alive. They had been for three days without food or water.

Two theatrical people were in a hotel in Santa Rosa when the shock came. The room was on the fourth floor. The roof collapsed. One of them was thrown from the bed and both were caught by the descending timbers and pinned helplessly beneath the debris. They could speak to each other and could touch one another's hands, but the weight was so great that they could do nothing to liberate themselves. After three hours rescuers came, cut a hole in the roof and both were released uninjured.

Even the docks were converted into hospitals in the stringent exigency of the occasion, about 100 patients being stretched on Folsom street dock at one time. In the evening tugs conveyed them to Goat Island, where they were lodged in the hospital. The docks from Howard Street to Folsom Street had been saved, the fire at this point not being permitted to creep farther east than Main Street. Another series of fatalities occurred, caused by the stampeding of a herd of cattle at Sixth and Folsom Streets. Three hundred of the panic-stricken animals ran amuck when they saw and felt the flames and charged wildly down the street, trampling

REFUGEES CAMPING UNDER SHELTER OF THE MINT.

The United States Mint was one of the few public buildings unharmed by either earthquake or flames. Many of the homeless took advantage of the grass plots surrounding it to erect their temporary tents, and even made their homes on the sidewalk.

VIEW OF THE RUINS LOOKING TOWARD VAN NESS AVENUE.

This photograph, taken at Van Ness Avenue, where a mile of magnificent mansions were blown up to stay the progress of the flames, gives an unparalleled picture of desolation. The ruins of the $7,000,000 City Hall are seen in the distance.

AVERTING FAMINE IN SAN FRANCISCO.

Handing out provisions to the destitute in one of San Francisco's parks after the fire and earthquake had driven 300,000 people from the residential parts of the city and left them homeless.

LOOKING UP CALIFORNIA STREET.
This stately thoroughfare was choked with debris by the earthquake shock and then
laid waste by fire. The magnificent $3,000,000 Fairmount Hotel at the
top of the hill was saved, though considerably damaged.

under foot all who were in the way. One man was gored through and through by a maddened bull. At least a dozen persons, it is said, were killed, though probably this is an overestimate. One observer tells us that "the first sight I saw was a man with blood streaming from his wounds, carrying a dead woman in his arms. He placed the body on the floor of the court at the Palace Hotel, and then told me he was the janitor of a big building. The first he knew of the catastrophe he found himself in the basement, his dead wife beside him. The building had simply split in two, and thrown them down."

In the camps of refuge the deaths came frequently. Physicians were everywhere in evidence, but, without medicine or instruments, were fearfully handicapped. Men staggered in from their herculean efforts at the fire lines, only to fall gasping on the grass. There was nothing to be done. Injured lay groaning. Tender hands were willing, but of water there was none. "Water, water, for God's sake get me some water," was the cry that struck into thousands of souls of San Francisco.

The list of dead was not confined to San Francisco, but extended to many of the neighboring towns, especially to Santa Rosa, where sixty were reported dead and a large number missing, and to the insane asylum in its vicinity, from the ruins of which a hundred or more of dead bodies were taken.

THE FREE USE OF RIFLES.

A citizen tells us that "in the early part of the evening, and while the twilight lasts, there is a good deal of trafficking up and down the sidewalks. Having finished their dinners of government provisions, cooked on the street or in the parks, the people promenade for half an hour or so. By half-past eight the town is closed tight. A rat scurrying in the street will bring a soldier's rifle to his shoulder. Any one not wearing a uniform or a Red Cross badge is a suspicious character and may be shot unless he halts at command. Even the men in uniform do well to stop still, for it is hard to tell a uniform in the half light thrown up by the burning town and the great shadows.

"Last night two of us ventured out on Van Ness Avenue a little late. There came up the noise of some kind of a shooting scrape far down the street. We hurried in that direction to see what was doing. An eighteen-year-old boy in a uniform barred the way, levelled his rifle and said in a peremptory way:

" 'Go home.'

"We took a course down the block, where an older soldier, more communicative but equally peremptory, informed us that we were trifling with our lives, news or no news.

" 'We've shot about 300 people for one thing or another,' he said. 'Now, dodge trouble. Git!' That ended the expedition."

THE LOSS IN WEALTH.

If we pass now from the record of the loss of lives to that of the destruction of wealth, the estimates exceed by far any fire losses recorded in history.

The truth is that when flames eat out the heart of a great city, devour its vast business establishments, storehouses and warehouses, sweep through its centres of opulence, destroy its wharves with their accumulation of goods, spread ruin and havoc everywhere, it is impossible at first to estimate the loss. Only gradually, as time goes on, is the true loss discovered, and never perhaps very

accurately, since the owners and the records of riches often disappear with the wealth itself. In regard to San Francisco, the
early estimate was that three-fourths of the city, valued at $500,-
000,000, was destroyed.

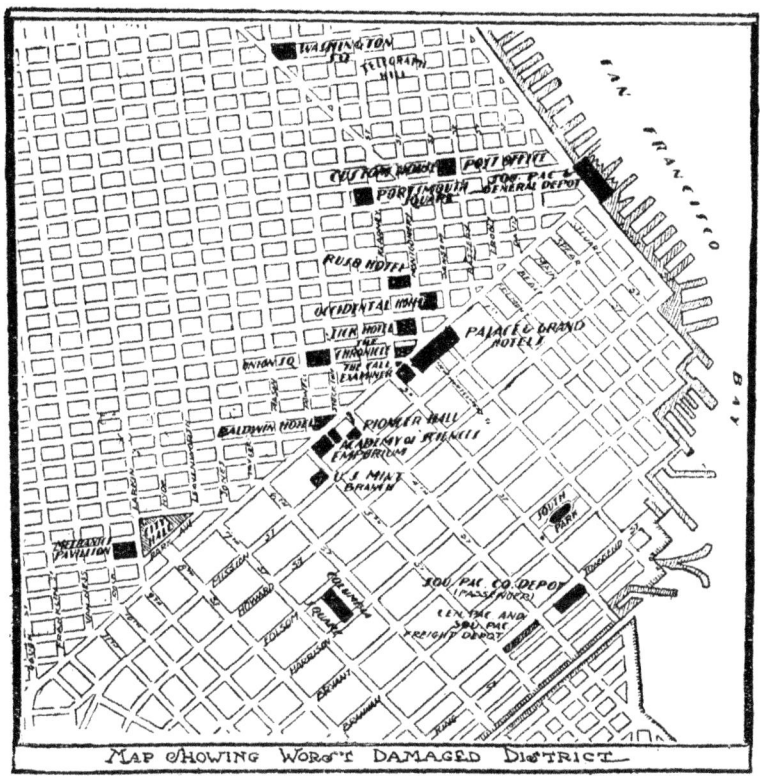

MAP SHOWING WORST DAMAGED DISTRICT

But early estimates are apt to be exaggerated, and on Friday,
two days after the disaster, we find this estimate reduced to $250,-
000,000. A few more days passed and these figures shrunk still
further, though it was still largely conjectural, the means of making
a trustworthy estimate being very restricted. Later on the pendulum

swung upward again, and two weeks after the fire the closest estimates that could be made fixed the property loss at close to $350,-000,000, or double that of the Chicago fire. But as the actual loss in the latter case proved considerably below the early estimates, the same may prove to be the case with San Francisco.

Special personal losses were in many cases great. Thus the Palace Hotel was built at a cost of $6,000,000, and the St. Francis, which originally cost $4,000,000, was being enlarged at great expense. Several of the great mansions on Nob's Hill cost a million or more, the City Hall was built at a cost of $7,000,000, the new Post Office was injured to the extent of half a million, while a large number of other buildings might be named whose value, with their contents, was measured in the millions.

It was not until May 3d that news came over the wires of another serious item of loss. The merchants had waited until then for their fire-proof safes and vaults to cool off before attempting to open them. When this was at length done the results proved disheartening. Out of 576 vaults and safes opened in the district east of Powell and north of Market Street, where the flames had raged with the greatest fury, it was found that fully forty per cent. had not performed their duty. When opened they were found to contain nothing but heaps of ashes. The valuable account books, papers and in some cases large sums of money had vanished, the loss of the accounts being a severe calamity in a business sense. As all the banks were equipped with the best fire-proof vaults, no fear was felt for the safety of their contents.

LOOTERS IN CHINATOWN.

Chinatown suffered severely, the merchants of that locality possessing large stocks of valuable goods, many of which were looted by seemingly respectable sightseers after the ruins had cooled off, bronze, porcelain and other valuable goods being taken from the ruins. One example consisted in a mass of gold and silver valued at $2,500, which had been melted by the fire in the store of Tai Sing, a Chinese merchant. This was found by the police on May 3d in a place where it had been hidden by looters.

But with all its losses San Francisco does not despair. The spirit of its citizens is heroic, and there are some hopeful signs in the air. The insurances due are estimated to approximate $175,-000,000, and there are other moneys likely to be spent on building during the coming year, making a total of over $200,000,000. Eastern capitalists also talk of investing $100,000,000 of new capital in the rebuilding of the city, while the San Francisco authorities have a project of issuing $200,000,000 of municipal bonds, the payment to be guaranteed by the United States Government. Thus, two weeks after the earthquake, daylight was already showing strongly ahead and hope was fast beginning to replace despair.

CHAPTER VIII.

Wonderful Record of Thrilling Escapes.

SHUDDERING under the memories of what seems more like a nightmare than actual reality to the survivors of this frightful calamity, they have tried to picture in words far from adequate the days of terror and the nights of horror that fell to the lot of the people of the Golden Gate city and their guests.

They recount the roar of falling structures and the groans and pitiful cries of those pinned beneath the timbers of collapsing buildings. They speak of their climbing over dead bodies heaped in the streets, and of following tortuous ways to find the only avenue of escape—the ferry, where men and women fought like infuriated animals, bent on escape from a fiery furnace.

These refugees tell of the great caravan composed of homeless persons in its wild flight to the hills for safety, and in that great procession women, harnessed to vehicles, trudging along and tugging at the shafts, hauling all that was left of their earthly belongings, and a little food that foresight told them would be necessary to stay the pangs of hunger in the hours of misery that must follow.

We give below an especially accurate picture from the description of the well-known writer, Jane Tingley, who, an eye-witness of it all, did so much to help the sufferers, and who, with all the unselfishness of true American womanhood, sacrificed her own comfort and needs for those of others.

GARDENS OF A BEAUTIFUL PLACE NEAR SANTA BARBARA.
The climate of California is world-renowned for its mildness, and thousands of villas dot
hills near the coast.

"May God be merciful to the women and children in this land of desolation and despair!" she wrote on April 21st.

"Men have done, are doing such deeds of sublime self-sacrifice, of magnificent heroism, that deserve to make the title of American manhood immortal in the pages of history. The rest lies with the Almighty.

"I spent all of last night and to-day in that horror city across the bay. I went from this unharmed city of plenty, blooming with abounding health, thronged with happy mothers and joyous children, and spent hours among the blackened ruins and out on the windswept slopes of the sand hills by the sea, and I heard the voice of Rachel weeping for her children in the wilderness and mourning because she found them not.

"I climbed to the top of Strawberry Hill, in Golden Gate Park, and saw a woman, half naked, almost starving, her hair dishevelled and an unnatural lustre in her eyes, her gaze fixed upon the waters in the distance, and her voice repeating over and over again: 'Here I am, my pretties; come here, come here.'

"I took her by the hand and led her down to the grass at the foot of the hill. A man—her husband—received her from me and wept as he said: 'She is calling our three little children. She thinks the sounds of the ocean waves are the voices of our lost darlings.'

"Ever since they became separated from their children in that first terrific onrush of the multitude when the fire swept along Mission Street these two had been tramping over the hills and parks without food or rest, searching for their little ones. To all whom they have met they have addressed the same pitiful question: 'Have you seen anything of our lost babies?' They will not know what has become of them until order has been brought out of chaos; until the registration headquarters of the military authorities has secured the names of all who are among the straggling wanderers around the camps of the homeless. Perhaps then it will be found that these children are in a trench among the corpses of the weaklings who have succumbed to the frightful rigors of the last three days.

"Last night a soldier seized me by the arm and cried: 'If you are a woman with a woman's heart, go in there and do whatever you can.'

" 'In there' meant behind a barricade of brush, covered with a blanket that had been hastily thrown together to form a rude shelter. I went in and saw one of my own sex lying on the bare grass naked, her clothing torn to shreds, scattered over the green beside her. She was moaning pitifully, and it needed no words to tell a woman what the matter was, I bade my man escort to find a doctor, or at least send more women at once. He ran off and soon two sympathetic ladies hastened into the shelter. In an hour my escort returned with a young medical student. Under the best ministrations we could find, a new life was ushered into this hell, which, a few hours before, was the fairest among cities.

" 'There have been many such cases,' said the medical student. "Many of the mothers have died—few of the babies have lived. I, personally, know of nine babies that have been born in the park to-day. There must have been many others here, among the sand hills, and at the Presidio."

"Think of it, you happy women who have become mothers in comfortable homes, attended with every care that loving hands can bestow. Think of the dreadful plight of these poor members of your sex. The very thought of it is enough to make the hearts of women burst with pity.

"To-day I walked among the people crowded on the Panhandle. Opposite the Lyon Street entrance, on the north side, I saw a young woman sitting tailor-fashion in the roadway, which, in happier days, was the carriage boulevard. She held a dishpan and was looking at her reflection in the polished bottom, while another girl was arranging her hair. I recognized a young wife, whose marriage to a prominent young lawyer eight months ago was a gala event among that little handful of people who clung to the old-time fashionable district of Valencia Street, like the Phelan and Dent families, and refused to move from that aristocratic section when the new-made millionaires began to build their palaces on Nob Hill and Pacific Heights. I spoke to the young woman about the disadvantages of making her toilet under such untoward circumstances.

" 'Ah, Julia, dear, you must stay to luncheon,' she said, extending her fingers just as though she stood in her own drawing-room.

MISERY DRIVES SOME INSANE.

"I looked at the maid in astonishment, for I had never met the young society woman before. The maid shook her head and whispered when she got the chance:

" 'My mistress is not in her right mind.'

" 'Where is her husband?' I asked.

" 'He has gone to try to get some food,' said the girl. 'She imagines that she is in her own home, before her dressing table, and is having me do up her hair against some of her friends dropping in.'

" 'She must have suffered,' I said, 'to cause such a mental derangement.

"The girl's eyes filled with tears. She told me that her mistress had seen her brother killed by falling timbers while they were hurrying to a place of safety. A little farther on I saw two women concealed as best they might be behind a tuft of sand brush, one lying face down on the ground, while the other vigorously massaged her bare back. I asked if I might help, and learned that the ministering angel was the unmarried daughter of one of the city's richest merchants, and that the girl whom she succored had been employed as a servant in her father's household. The girl's back had been injured by a fall, and her mistress' fair hands were trying to make her well again.

"Thus has this overwhelming common woe levelled all barriers of caste and placed the suffering multitude on a basis of democracy. On a rock behind a manzanita bush near the edge of Stow Lake I saw a Chinaman making a pile of broken twigs in the early morning. The man felt inside his blouse and swore a gibbering, unintelligible Asiatic oath as his hand came forth empty. Observing my escort, the Chinaman approached and said:

" 'Bosse, alle same, catchee match?'

"My escort gave him the desired article, and the Chinaman made a fire of his pile of twigs. 'Why are you making a fire, John?' I asked.

" 'Bleakfast,' he replied laconically.

"I asked him where his food might be, and be gave us a quick glance of suspicion as he said briefly, 'No sabbe.'

"We stood watching him, evidently to his great distress, and finally he made bold to say, 'You no stand lound, bosse. You go 'way.'

"We left him, but after making the tour around the lake came back to the same place. There sat four people on the ground eating fried pork, potatoes and Chinese cakes. In a young woman of the group I recognized one whom I had seen dancing at one of Mr. Greenway's Friday Night Cotillion balls in the Palace Hotel's maple room during the winter. They offered to share their meal with us, but we told them that we had just come from breakfast in Oakland. I told them about the strange conduct of their Chinaman, who was traveling back and forth from his fire to the 'table' with the food as it became ready to serve.

"The father of the family laughed.

SOCIETY FOLKS COMPELLED TO CAMP.

" 'Yes,' he said, 'that is Charlie's way. He has been with us many years, and when our home was destroyed he came out here with us in preference to seeking refuge among his countrymen in Chinatown. Yesterday we were without food, and Charlie disappeared. I thought he had deserted us, but toward dark he came back with a bamboo pole over his shoulder and a Chinese market gardener's basket suspended from either end. In one of the baskets he had a pile of blankets and a lot of canvas. In the other was an assortment of pork, flour, Chinese cakes and vegetables, besides a half-dozen chickens and a couple of bagfuls of rice.

" 'Charlie had been foraging in Chinatown for us before the fire reached that quarter. He made a tent and improvised beds for us, and he has the food concealed somewhere in the vicinity, but where he will not tell us, for fear that we will give some of it to others and reduce our own supply. Charlie boils rice for himself.

He will not touch the other food. Without him we should have been starving.'"

G. A. Raymond, who was in the Palace Hotel when the earthquake occurred, says:

"I had $600 in gold under my pillow. I awoke as I was thrown out of bed. Attempting to walk, the floor shook so that I fell. I grabbed my clothing and rushed down into the office, where dozens were already congregated. Suddenly the lights went out, and every one rushed for the door.

"Outside I witnessed a sight I never want to see again. It was dawn and light. I looked up. The air was filled with falling stones. People around me were crushed to death on all sides. All around the huge buildings were shaking and waving. Every moment there were reports like 100 cannon going off at one time. Then streams of fire would shoot out, and other reports followed.

"I asked a man standing by me what had happened. Before he could answer a thousand bricks fell on him and he was killed. A woman threw her arms around my neck. I pushed her away and fled. All around me buildings were rocking and flames shooting. As I ran people on all sides were crying, praying and calling for help. I thought the end of the world had come.

"I met a Catholic priest, and he said: 'We must get to the ferry.' He knew the way, and we rushed down Market Street. Men, women and children were crawling from the debris. Hundreds were rushing down the street, and every minute people were felled by falling debris.

"At places the streets had cracked and opened. Chasms extended in all directions. I saw a drove of cattle, wild with fright, rushing up Market Street. I crouched beside a swaying building. As they came nearer they disappeared, seeming to drop into the earth. When the last had gone I went nearer and found they had indeed been precipitated into the earth, a wide fissure having swallowed them. I worked my way around them and ran out to the ferry. I was crazy with fear and the horrible sights,

"How I reached the ferry I cannot say. It was bedlam, pandemonium and hell rolled into one. There must have been 10,000 people trying to get on that boat. Men and women fought like wild cats to push their way aboard. Clothes were torn from the backs of men and women and children indiscriminately. Women fainted, and there was no water at hand with which to revive them. Men lost their reason at those awful moments. One big, strong man, beat his head against one of the iron pillars on the dock, and cried out in a loud voice: 'This fire must be put out! The city must be saved!' It was awful.

TERRIBLE SCENE AT THE FERRY.

"When the gates were opened the mad rush began. All were swept aboard in an irresistible tide. We were jammed on the deck like sardines in a box. No one cared. At last the boat pulled out. Men and women were still jumping for it, only to fall into the water and probably drown."

The members of the Metropolitan Opera Company, of New York, were in San Francisco at this time, and nearly all of these famous singers, known all over the world, suffered from the great disaster.

All of the splendid scenery, stage fittings, costumes and musical instruments were lost in the fire, which destroyed the Grand Opera House, where the season had just opened to splendid audiences.

Many of the operatic stars have given very interesting accounts of their experiences. Signor Caruso, the famous tenor and one of the principals of the company, had one of the most thrilling experiences. He and Signor Rossi, a favorite basso, and his inseparable companion, had a suite on the seventh floor and were awakened by the terrific shaking of the building. The shock nearly threw Caruso out of bed. He said:

"I threw open the window, and I think I let out the grandest notes I ever hit in all my life. I do not know why I did this. I presume I was too excited to do anything else.

GREAT SINGERS ESCAPE.

"Looking out of the window, I saw buildings all around rocking like the devil had hold of them. I wondered what was going on. Then I heard Rossi come scampering into my room. 'My God, it's an earthquake!' he yelled. 'Get your things and run!' I grabbed what I could lay my hands on and raced like a madman for the office. On the way down I shouted as loud as I could so the others would wake up.

"When I got to the office I thought of my costumes and sent my valet, Martino, back after them. He packed things up and carried the trunks down on his back. I helped him take them to Union Square."

It is said that ten minutes later he was seen seated on his valise in the middle of the street. But to continue his story:

"I walked a few feet away to see how to get out, and when I came back four Chinamen were lugging my trunks away. I grabbed one of them by the ears, and the others jumped on me. I took out my revolver and pointed it at them. They spit at me. I was mad, but I hated to kill them, so I found a soldier, and he made them give up the trunks.

"Ah, that soldier was a fine fellow. He went up to the Chinamen and slapped them upon the face, once, twice, three times. They all howled like the devil and ran away. I put my revolver back into my pocket, and then I thanked the soldier. He said: " 'Don't mention it. Them Chinks would steal the money off a dead man's eyes.' "

They say that Rossi, though almost in tears, was heard trying his voice at a corner near the Palace Hotel.

TEDDY'S PICTURE PROVES "OPEN SESAME."

"I went to Lafayette Square and slept on the grass. When I tried to get into the square the soldiers pushed me back. I pleaded with them, but they would not listen. I had under my arm a large photograph of Theodore Roosevelt, upon which was written: 'With kindest regards from Theodore Roosevelt.' I showed them this, and one of them said: 'If you are a friend of Teddy, come in and make yourself at home.'

"I put my trunks in the cellar of the Hotel St. Francis and thought they would be safe. The hotel caught fire, and my trunks were all burned up. To think I took so much trouble to save them!"

In spite of the news of all the woe and suffering which we hear, it is cheering to learn also of the many thousands of heroic deeds by brave men during the terrible scenes enacted through the four days passing since the eventful morning when the earth began to demolish splendid buildings of business and residence and fire sprang up to complete the city's destruction. The Mayor and his forces of police, the troops under command of General Funston, volunteer aids to all these, and the husbands of terrified wives, and the sons, brothers and other relatives who toiled for many consecutive hours through smoke and falling walls and an inferno of flames and explosions and traps of danger of all kinds, often without food or water—toiling as men never toiled before to save life and relieve distress of all kinds—all these were examples of heroism and devotion to duty seldom witnessed in any scenes of terror in all time. There are brave, unselfish men and heroic women yet in the world, and all of the best of human nature has been exhibited in large dimensions in the terrible disaster at San Francisco.

CHAPTER IX.

Disaster Spreads Over the Golden State

THE first news that the world received of the earthquake came direct from San Francisco and was confined largely to descriptions of the disaster which had overwhelmed that city. It was so sudden, so appalling, so tragic in its nature, that for the time being it quite overshadowed the havoc and misery wrought in a number of other California towns of lesser note.

As the truth, however, became gradually sifted out of the tangle of rumors, the horror, instead of being diminished, was vastly increased. It became evident that instead of this being a local catastrophe, the full force of the seismic waves had travelled from Ukiah in the north to Monterey in the south, a distance of about 180 miles, and had made itself felt for a considerable distance from the Pacific westward, wrecking the larger buildings of every town in its path, rending and ruining as it went, and doing millions of dollars worth of damage.

THE DESTRUCTION OF SANTA ROSA.

In Santa Rosa, sixty miles to the north of San Francisco, and one of the most beautiful towns of California, practically every building was destroyed or badly damaged. The brick and stone business blocks, together with the public buildings, were thrown down. The Court House, Hall of Records, the Occidental and Santa Rosa Hotels. the Athenæum Theatre, the new Masonic Temple, Odd Fellows' Block, all the banks, everything went, and in all the city not one brick or stone building was left standing, except the California Northwestern Depot.

In the residential portion of the city the foundations receded from under the houses, badly wrecking about twenty of the largest and damaging every one more or less; and here, as in San Francisco, flames followed the earthquake, breaking out in a dozen different places at once and completing the work of devastation. From the ruins of the fallen houses fifty-eight bodies were taken out and interred during the first few days, and the total of dead and injured was close to a hundred. The money loss at this small city is estimated at $3,000,000.

The destruction of Santa Rosa gave rise to general sorrow among the residents of the interior of the State. It was one of the show towns of California, and not only one of the most prosperous cities in the fine county of Sonoma, but one of the most picturesque in the State. Surrounding it there were miles of orchards, vineyards and corn fields. The beautiful drives of the city were adorned with bowers of roses, which everywhere were seen growing about the homes of the people. In its vicinity are the famous gardens of Luther Burbank, the "California wizard," but these fortunately escaped injury.

At San Jose, another very beautiful city of over 20,000 population, not a single brick or stone building of two stories or over was left standing. Among those wrecked were the Hall of Justice, just completed at a cost of $300,000; the new High School, the Presbyterian Church and St. Patrick's Cathedral. Numbers of people were caught in the ruins and maimed or killed. The death list appears to have been small, but the property damage was not less than $5,000,000. The Agnew State Insane Asylum, in the vicinity of San Jose, was entirely destroyed, more than half the inmates being killed or injured.

THE STANFORD UNIVERSITY.

The Leland Stanford, Jr., University, at Palo Alto (about thirty miles south of San Francisco), felt the full force of the earthquake and was badly wrecked. Only two lives were lost as a result of the earthquake, one of a student, the other of a fireman, but eight students were injured more or less seriously. The damage to the buildings is estimated by President Jordan to amount to about $4,000,000.

The memorial church, with its twelve marble figures of the apostles, each weighing two tons, was badly injured by the fall of its Gothic spire, which crashed through the roof and demolished much of the interior; the great entrance archway was split in twain and wrecked; so, too, were the library, the gymnasium and the power house. A number of other buildings in the outer quadrangle and some of the small workshops were seriously damaged.

Encina Hall and the inner quadrangle were practically uninjured, and the bulk of the books, collections and apparatus escaped damage.

Sacramento, together with all the smaller cities and towns that dot the great Sacramento Valley for a distance fifty miles south and 150 miles north of the capital, escaped without injury, not a single pane of glass being broken or a brick displaced in Sacramento and no injury done in the other places, they lying eastward of the seat of serious earthquake activity.

Los Angeles and Santa Barbara escaped with a slight trembling; Stockton, 103 miles north of San Francisco, felt a severe shock and the Santa Fe bridge over the San Joaquin River at this point settled several inches. The only place in Southern California that suffered was Brawley, a small town lying 120 miles south of Los Angeles, about 100 buildings in the town and the surrounding valley being injured, though none of them were destroyed.

THE EARTHQUAKE AT OTHER CITIES.

At Alameda, on the bay opposite San Francisco, a score of chimneys were shaken down and other injuries done. Railroad tracks were twisted, and over 600 feet of track of the Oakland Transit Company's railway sank four feet. The total damage done amounted to probably $200,000, but no lives were lost. Tomales, a place of 350 inhabitants, was left a pile of ruins.

At Los Panos several buildings were wrecked, causing damage to the extent of $75,000, but no lives were lost.

At Loma Prieta the earthquake caused a mine house to slip down the side of a mountain, ten men being buried in the ruins.

Fort Bragg, one of the principal lumbering towns in Mendocino County, was practically wiped out by fire following the earthquake, but out of a population of 5,000 only one was killed, though scores were injured.

The town of Berkeley, across the bay from San Francisco, suffered considerable damage from twisted structures, fallen walls and broken chimneys, the greatest injury being in the collapse of the town hall and the ruin of the deaf and dumb asylum. The University of California, situated here, was fortunate in escaping injury, it being reported that not a building was harmed in the slightest degree. Another public edifice of importance and interest, in a different section of the State, the famous Lick Astronomical Observatory, was equally fortunate, no damage being done to the buildings or the instruments.

AT THE STATE UNIVERSITY.

Salinas, a town down the coast near Monterey, suffered severely, the place being to a large extent destroyed, with an estimated loss of over $1,000,000. The Spreckels' sugar factory and a score of other buildings were reported ruined and a number of lives lost. During the succeeding week several other shocks of some strength were reported from this town.

Thus the ruinous work of the earthquake stretched over a broad track of prosperous, peaceful and happy country, embracing one of the best sections of California, laying waste not only the towns in its path, but doing much damage to ranch houses and country residences. Strange manifestations of nature were reported from the interior, where the ground was opened in many places like a ploughed field. Great rents in the earth were reported, and for many miles north from Los Angeles miniature geysers are said to have spouted volcano-like streams of hot mud.

Railroad tracks in some localities were badly injured, sinking or lifting, and being put out of service until repaired. In fact, the ruinous effects of the earthquake immensely exceeded those of any similar catastrophe ever before known in the United States, and when the destruction done by the succeeding conflagration in San Francisco is taken into account the California earthquake of 1906 takes rank with the most destructive of those recorded in history.

CHAPTER X.

All America and Canada to the Rescue

URING the first three days after the terrible news had been
flashed over the world the relief fund from the nation had
leaped beyond the $5,000,000 mark. New York took the
lead in the most generous giving that the world has ever seen.
From every town and country village the people hastened to the
Town Halls, the newspaper offices and wherever help was to be
found most quickly, to add their savings and to sacrifice all but
necessities for their stricken fellow-countrymen. Never has there
been such a practical illustration of brotherly love. A perfect
shower of gold and food was poured out to the sufferers to give
them immediate assistance and to help them to a new start in life.
All relief records were broken within two days of the disaster, but
still the purses of the rich and poor alike continued to add to the
huge contributions. Though the relief records were broken, every
succeeding dispatch from the West told too plainly the terrible
fact that all records of necessity were also broken.

Over the entire globe Americans wherever they were hastened to cable or telegraph their bankers to add their share to the
great work. A large fund was at once started in London, and with
contributions of from $2,000 to $12,000 the sum was soon raised
to hundreds of thousands of dollars.

Individual contributions of $100,000 were common.   In addition to John D. Rockefeller's gift of this sum, his company, the Standard Oil, gave another $100,000.   The Steel Corporation and Andrew Carnegie each gave $100,000.   From London William Waldorf Astor cabled his American representative, Charles A. Peabody, to place $100,000 at once at the disposal of Mayor Schmitz, of San Francisco, which was done.   The Dominion Government of Canada made a special appropriation of $100,000 and the Canadian Bank of Commerce, at Toronto, gave $10,000.   And two of the great steamship companies owned in Germany sent $25,000 each.

RIGHT OF WAY FOR FOOD TRAINS.

On nearly a dozen roads, two days before the fire was over, great trains of freight cars loaded with foodstuffs were hastening at express speed to San Francisco.   They had the right of way on every line.   E. H. Harriman, in addition to giving $200,000 for the Union Pacific, Southern Pacific and other Harriman roads, issued orders that all relief trains bound for the desolated city should have precedence over all other business of the roads.

Advices from many points indicated that at least 150 freight cars loaded with the necessaries so eagerly awaited in San Francisco were speeding there as fast as steam could drive them.   In addition, several steamers from other Pacific coast points, all food-laden, were rushing toward the stricken city.

The rapidity with which the various relief funds in every city grew was almost magical.

From corporations, firms, labor unions, religious societies, individuals, rich and poor, money flowed. Even the children in the schools gave their pennies. Every grade of society, every branch of trade and commerce seemed inspired by a spirit of emulation in giving.

The United States Government at once voted a contribution of $1,000,000, and government supplies were rushed from every post in the West.

The $1,000,000 government gift, which formed the nucleus of the relief fund, was doubled on Saturday by a resolution appropriating another, and a vote was taken on Monday to increase this sum to $1,500,000, making a total government contribution of $2,500,000. This was largely expended in supplies of absolute necessaries, furnished from the stores of the War Department, and those first sent being five carloads of army medical supplies from St. Louis. A cargo of evaporated cream was also sent to use in the care of little children, while the Red Cross Society shipped a carload of eggs from Chicago. Dr. Edward Devine, special Red Cross agent in San Francisco, was appointed to distribute these supplies.

CARGOES OF SUPPLIES.

Trainloads of other supplies were dispatched in all haste from various points in the West and East, carrying provisions of all kinds, tents, cots, clothing, bedding and a great variety of other articles. A special train of twenty-six cars was dispatched from Portland, Oregon, on Thursday night, conveying ten doctors, twenty trained nurses and 800,000 pounds of provisions. Chicago sent meat. Minneapolis sent flour, and, in fact, every part of the country moved in the greatest haste for the relief of the stricken city.

There was urgent need of haste. On Friday, while the flames were still making their way onward, General Funston telegraphed: "Famine seems inevitable." The people of the country took a more hopeful view of it, and by Saturday night the spectre of famine was definitely driven from the field and food for all the fugitives was within reach.

THE SYMPATHY OF THE PEOPLE AWAKES.

On all sides the people were awake and doing. In all the great cities agencies to receive contributions were opened, and many of the newspapers undertook the task of collecting and forwarding supplies. The smaller towns were equally alert in furnishing their quota to the good work, and from countryside and village contributions were forwarded until the fund accumulated to an unprecedented amount. Collections were made in factories, in stores, in offices, in the public schools; cash boxes or globes stood in all frequented places and were rapidly filled with bank notes; theatrical and musical entertainments were given for the benefit of the earthquake sufferers; never had there been such an awakening. As an instance of the spirit displayed, one man came running into a banking house and threw a thousand dollar bill on the counter.

"For San Francisco," he said, as he turned toward the door.

"What name?" asked the teller.

"Put it down to 'cash,' " he answered, as he vanished.

Rapidly the fund accumulated. A few days brought it up to

the $5,000,000 mark. Then it grew to $10,000,000. Within ten days' time the relief fund was estimated at $18,000,000, and the good work was still going on—in less profusion, it is true, but still the spirit was alive.

FOREIGN OFFERS OF AID.

The generous impulse was not confined to the United States. From all countries came offers of aid. Canada was promptly in the field, and the chief nations of Europe were quick to follow, while Japan made a generous offer, and in far Australia funds were started at the various cities for the sufferers. No doubt a large sum from foreign lands would have been available had not President Roosevelt declined to accept contributions from abroad, as not needed in view of America's abundant response. To the Hamburg-Line, which offered $25,000, the following letter was sent:

"The President deeply appreciates your message of sympathy, and desires me to thank you heartily for the kind offer of outside aid. Although declining, the President earnestly wishes you to understand how much he appreciates your cordial and generous sympathy."

All other offerings from abroad were in the same thankful spirit declined, even those from our immediate neighbors, Canada and Mexico. Some feeling was aroused by this, especially in the relief committee at San Francisco, which felt that the need of that city was so great and urgent that no offer of relief should have been declined. In response the President explained that he only spoke for the government, in his official capacity, and that San Francisco was in no sense debarred from accepting any contributions made directly to it.

It may justly be said for the people of this country that their spontaneous generosity in the presence of a great calamity, either at home or abroad, is always magnificent. It never waits for solicitation. It does not delay even until the necessity is demonstrated, but it assumes that where there is great destruction of property and homes are swept away there must be distress which calls for immediate relief.

There is one ray of light in the gloom caused by the calamity at San Francisco. A truly splendid display of brotherly love and sympathy has been shown by the people of this country, and a similar display was ready to be shown by the people of the civilized world had it been felt that the occasion demanded it and that the exigency surpassed the power of our people to meet it.

ENTERPRISE IN SAN FRANCISCO.

In the face of an appalling and death-dealing disaster, rendering an entire community dependent for the bare necessities of life and putting it in imminent danger of greater horrors, the nation has been stirred as it has rarely been before, and there have been awakened those deeper feelings of brotherhood which are referred to in the oft-quoted passage that "one touch of nature makes the whole world akin."

The nature indicated in this instance is human nature in its highest manifestation, the sympathetic sentiment that stirs deeply in all our hearts and needs but the occasion to make itself warmly manifested. There is something incomparably splendid in the spectacle of an entire nation straining every nerve to send succor to the helpless and the suffering, and this spectacle has warmed the hearts of our people to the uttermost and inspired them to make the most strenuous efforts to drive away the gaunt wolf of famine from the ruined homes of our far Pacific brethren.

It may be said that San Francisco will be willing to accept this relief only so long as stern necessity demands it. At this writing only two weeks have passed since the dread calamity, and already active steps are being taken to provide for themselves. As an example of their enterprise, it may be said that their newspapers hardly suspended at all, the *Evening Post* alone suspending publication for a time from being unable to acquire a plant in the vicinity of the city. When the conflagration made it apparent that all plants would be destroyed, the *Bulletin* put at work a force in its composing rooms, a hand-bill was set and some hundreds of copies run off on the proof-press, giving the salient features of the day's news.

The morning papers, the *Call, Chronicle* and *Examiner,* retired to Oakland, on the other side of the bay, and there, on Thursday morning, issued a joint paper from the office of the *Oakland Tribune.* On Friday morning they split forces again, the *Examiner* retaining the use of the *Tribune* plant and the *Call* and *Chronicle* issuing from the office of the *Oakland Herald.* Two days later the *Call* secured the service of the *Oakland Enquirer* plant. Meantime, on Friday, the *Bulletin,* after a suspension of one day, made arrangements for the use in the afternoon of the *Oakland Herald* equipment, and from these sources and under such circumstances the San Francisco papers have been issuing.

Offices were hurriedly opened on Fillmore Street, which today is the main thoroughfare of San Francisco, and from these headquarters the news of the day as it is gathered is transmitted by means of automobiles and ferry service to the Oakland shore.

There also were accepted such advertisements as had been offered. The number of these was, perhaps, the best visual sign of the resurrection of the new city. It was noted that in a fourteen-page paper printed within two weeks after the fire by the *Examiner* there were over nine pages of advertisements, and in a sixteen-page paper published by the *Chronicle* at least fifty per cent. of its space was devoted to the same end.

Many of the larger factories left unharmed were also quick to start work. At the Union Iron Works 2,300 men were promptly employed, and the management expected within a fortnight to have the full complement of its force, nearly 4,000 men, engaged. No damage was done to the three new warships being built at these works for the government, the cruisers California and Milwaukee and the battleship South Dakota. The steamer City of Puebla, which was sunk in the bay, has been raised and is being repaired. Workmen are also engaged fixing the steamship Columbia, which was turned on her side. The hulls of the new Hawaiian-American Steamship Company's liners were pitched about four feet to the south, but were uninjured and only need to be replaced in position.

As for the working people at large, those without funds for their own support, abundant employment will quickly be provided for them in the necessary work of clearing away the debris, thus opening the way to a resumption of business and reducing the number requiring relief. The ukase has already been issued that all able-bodied men needing aid must go to work or leave the city.

This dictum of Chief of Police Dinan's will be strictly enforced. The relief work and distribution of food and clothing are attracting a certain element to the city which does not desire to labor, while some already here prefer to live on the generosity of others. Chief

Dinan has determined that those who apply for relief and refuse work when it is offered them shall leave the city or be arrested for vagrancy. The police judges have suggested establishing a chain gang and putting all vagrants and petty offenders at work clearing up the ruins.

Perhaps never in the history of the city has there been so little crime in San Francisco. With the saloons closed, Chinatown, the Barbary Coast, and other haunts of criminals wiped out, and soldiers and marines on almost every block in the residence districts, there have been few crimes of any kind. It is the opinion of the police that most of the criminal element has left the city. The saloons, in all probability will remain closed for two more months.

THE PROBLEM OF THE CHINESE.

In conclusion of this chapter it is advisable to refer to the situation of one of the elements of San Francisco's population, the people of Chinatown. One of the problems facing the relief committees on both sides of the bay is the sheltering of the Chinese. Many of them are destitute. It has long been a question in San Francisco what should be done with Chinatown, and moving the Chinese in the direction of Colma has been agitated. Now they are without homes and without prospects of procuring any. They can get no land. The limits of Oakland's Chinatown have already been extended, and the strictest police regulations are in force to prevent further enlargement. On this side of the bay they are camping in open lots. Unless the government undertakes their relief, they are in grave danger. Those who have money cannot purchase property, as no one will sell to them. Few, however, even of the wealthiest merchants in Chinatown, saved anything of value, for their wealth was invested in the Oriental village which had sprung up in the heart of the area burned.

Yet it is the desire of the municipality not to harass this portion of its foreign population, and the vexatious problem of placing the new Chinatown will probably be settled to the satisfaction of the Chinese colony. This colony diverts an important part of the trade of San Francisco to that city, and if its members are dealt with unjustly there is danger of losing this trade. The question is one that must be left for the future to decide, but no doubt care will be taken that a new Chinatown with the unsavory conditions of the old shall not arise.

CHAPTER XI.

San Francisco of the Past

THE story of San Francisco's history and tragedy appeal with extraordinary force to the imagination of all civilized men. For several generations the city was looked upon as an Arabian Night's dream—a place where gold lay in the streets and joy and happiness were unlimited. Its settlement, or, rather, its real rise as a city, was as by magic. It was first a city of tents, of shanties, of "shacks," lying on the rim of a great, spacious bay. Ships of all sizes and rigs brought gold-seekers and provisions from the East, all the way round Cape Horn, after voyages of weary months, and at San Francisco their crews deserted and hundreds of these craft were left at their moorings to rot. Ashore was a riot of money, prodigious extravagance, mean, shabby appointments, sudden riches, great disappointment, revelry, improvidence and suicide.

The streets that now lay squares from the water were then at the water's edge and batteaus brought cargoes ashore. Long wharves—one was for years called the Long Wharf even after there were others built much longer—led out over the shallow water. These shallows were later filled and streets built upon them, and upon them arose warehouses, hotels, factories, lodging houses and business places.

The city grew rapidly in the direction away from the bay. But in its early days it was a city with no confidence in its own stability,

and its buildings were accordingly unstable. A few minor earthquakes shook some of these down years ago and established in the
minds of the people a horror of earthquakes. Frame houses became
the rule.

In its ensuing life San Francisco developed the attributes of a
city of gayety tempered by business. The population, for the most
part, affected light-hearted scorn of money, or, rather, of saving
money. It made mirth of life, habituated itself to expect windfalls
such as miners and prospectors dream of, developed a moderate
amount of business, and enjoyed the day while there was sunlight
and the night when there was artificial light. The windfalls grew
less frequent, mining became a costly and scientific process, and
agriculture succeeded it. But, though it was only necessary to tickle
the land with a hoe and pour water upon the tickled spot, to have
it laugh with two, three or even four harvests a year, agriculturists
continued scarce. The Chinese truck farms, some of which lay
within the city's lines, supplied the small fruits and vegetables.
Across the bay white men farmed, and grapes, fruits, vegetables
and flowers of prodigious variety and monstrous dimensions were
grown. But Eastern men came to do the farming. The Californian
who himself was an "Argonaut," or whose father was an Argonaut,
found no attractions in the steady labor of farming.

There followed a period of depression, ascribed by many to
the influx of the Chinese and their effect upon the labor market,
though the army of the unemployed were as a rule unwilling to do
the work their Celestial rivals engaged in, that of truck farming,
fruit raising, manual household labor, wood cutting and the like.
A heavy weight settled on the city; business grew slack; the army
of the unemployed, of ruined speculators and moneyless newcomers

grew steadily greater, and for an era San Francisco saw its dark side.

But this was not a long duration. There was fast developing a new and important business, resulting from the development of the real resources of the State—the fruits, particularly the citrous fruits that grew abundantly in the warm valley. Fortunes were made in oranges, lemons, limes, grapes, almonds and pears. Raisins, whose size defied anything heretofore known, were made from the huge grapes that grew in the San Joaquin Valley. Sonoma sent its grapes to be made into wine. Capital flowed in from every side. Eastern men in search of health, others in search of wealth, came to the Golden State. No matter who came, where they came from, or where they were going, they spent a few days, or many, and some money, or much, in " 'Frisco." The enterprise of the second edition pioneers quickly transformed the State and city.

AGRICULTURE BRINGS NEW WEALTH.

Luxury was startling. San Francisco's mercantile community equaled the best, the stores and shops were as beautiful as anywhere in the world and proportionately as well patronized. Theatres, music halls, restaurants, hotel bars and the like were ablaze with lights at night, and patronized by a gay throng. Sutro's bath, near the Cliff House, was a species of entertainment unequaled anywhere. The Presidio, as the army post is still known, as in the Spanish nomenclature, gave its drills, regarded as free exhibitions for the people. Golden Gate Park was an endless daily picnic ground.

The crowds in the streets of San Francisco were noticeably well dressed and usually gay, without that fixed, drawn, saturnine look noticeable among the people of the East. It is doubtful whether, upon the whole, the earnings of the San Francisco man equaled those of his Eastern brother, but his holidays were frequent and his joys greater. The grind of life was not yet steady—men had not become mere machines.

The climate of California is peculiar; it is hard to give an impression of it. In the first place, all the forces of nature work on laws of their own in that part of California. There is no thunder or lightning; there is no snow, except a flurry once in five or six years; there are perhaps half a dozen nights in the winter when the thermometer drops low enough so that there is a little film of ice on exposed water in the morning. Neither is there any hot weather. Yet most Easterners remaining in San Francisco for a few days remember that they were always chilly.

A PECULIAR YET DELIGHTFUL CLIMATE.

For the Gate is a big funnel, drawing in the winds and the mists which cool off the great, hot interior valley of San Joaquin and Sacramento. So the west wind blows steadily ten months of the year and almost all the mornings are foggy. This keeps the temperature steady at about 55 degrees—a little cool for comfort of an unacclimated person, especially indoors. Californians, used to it, hardly ever think of making fires in their houses except in the few exceptional days of the winter season, and then they rely mainly upon fireplaces. This is like the custom of the Venetians and the Florentines.

But give an Easterner six months of it, and he, too, learns to exist without a chill in a steady temperature a little lower than that to which he is accustomed at home. After that one goes about with perfect indifference to the temperature. Summer and winter San Francisco women wear light tailor-made clothes, and men wear the same fall-weight suits all the year around.

Except for the modern buildings, the fruit of the last ten years, the town presented at first sight to the newcomer a disreputable appearance. Most of the buildings were low and of wood. In the middle period of the 70's, when a great part of San Francisco was building, there was some atrocious architecture perpetrated. In that time, too, every one put bow windows on his house, to catch all of the morning sunlight that was coming through the fog, and those little houses, with bow windows and fancy work all down their fronts, were characteristic of the middle class residence districts.

Then the Italians, who tumbled over Telegraph Hill, had built as they listed and with little regard for streets, and their houses hung crazily on a side hill which was little less than a precipice. For the most part the Chinese, although they occupied an abandoned business district, had remade the houses Chinese fashion, and the Mexicans and Spaniards had added to their houses those little balconies without which life is not life to a Spaniard.

The hills are steep beyond conception. Where Vallejo Street ran up Russian Hill it progressed for four blocks by regular steps like a flight of stairs.

With these hills, with the strangeness of the architecture, and with the green gray tinge over everything, the city fell always into vistas and pictures, a setting for the romance which hung over everything, which has always hung over life in San Francisco since the padres came and gathered the Indians about Mission Dolores.

And it was a city of romance and a gateway to adventure. It opened out on the mysterious Pacific, the untamed ocean, and most of China, Japan, the South Sea Islands, Lower California, the west coast of Central America, Australia that came to this country passed in through the Golden Gate. There was a sprinkling, too, of Alaska and Siberia. From his windows on Russian Hill one saw always something strange and suggestive creeping through the mists of the bay. It would be a South Sea Island brig, bringing in copra, to take out cottons and idols; a Chinese junk with fan-like sails, back from an expedition after sharks' livers; an old whaler, which seemed to drip oil, back from a year of cruising in the Arctic. Even the tramp windjammers were deep-chested craft, capable of rounding the Horn or of circumnavigating the globe; and they came in streaked and picturesque from their long voyaging.

A MIXTURE OF RACES.

In the orange colored dawn which always comes through the mists of that bay, the fishing fleet would crawl in under triangular lateen sails, for the fishermen of San Francisco Bay are all Neapolitans, who have brought their costumes and sail with lateen rigs shaped like the ear of a horse when the wind fills them and stained an orange brown.

The "smelting pot of the races" Stevenson called the region along the water front, for here the people of all these craft met, Italians, Greeks, Russians, Lascars, Kanakas, Alaska Indians, black Gilbert Islanders, Spanish-Americans, wanderers and sailors from all the world, who came in and out from among the queer craft to lose themselves in the disreputable shanties and saloons. The

Barbary Coast was a veritable bit of Satan's realm. The place was made up of three solid blocks of dance halls, for the delectation of the sailors of the world. Within those streets of peril the respectable never set foot; behind the swinging doors of those saloons anything might be happening, crime was as common here as drink, and much went on of which the law was blankly ignorant.

Not far removed from this haunt of crime was the world-famous Chinatown, a district six blocks long and two wide, and housing when at its fullest some 30,000 Chinese. Old business houses at first, the new inmates added to them, rebuilt them, ran out their own balconies and entrances, and gave them that feeling of huddled irregularity which makes all Chinese built dwellings fall naturally into pictures. Not only this, they burrowed to a depth equal to three stories under the ground, and through this ran passages in which the Chinese transacted their dark and devious affairs —as the smuggling of opium, the traffic in slave girls and the settlement of their difficulties, by murder if they saw fit. The law was powerless to prevent or discover and convict the murderers.

Chinatown is gone; the Barbary Coast is gone; the haunts of crime have been swept by the devouring flames, and if the citizens can prevent they will never be restored. The old San Francisco is dead. The gayest, lightest-hearted, most pleasure-loving city of this continent, and in many ways the most interesting and romantic, is a horde of huddled refugees living among ruins. It may rebuild; it probably will; but those who have known that peculiar city by the Golden Gate and have caught its flavor of the Arabian Nights feel that it can never be the same. When it rises out of its ashes it will probably doubtless resemble other modern cities and have lost its old strange flavor.

CHAPTER XII.

Life in the Metropolis of the Pacific

B ROUGHT up in a bountiful country, where no one really has to work very hard to live, nurtured on adventure, scion of a free and merry stock, the real, native Californian is a distinctive type; as far from the Easterner in psychology as the extreme Southern is from the Yankee. He is easy going, witty, hospitable, lovable, inclined to be unmoral rather than immoral in his personal habits, and above all easy to meet and to know.

Above all there is an art sense all through the populace which sets it off from any other part of the country. This sense is almost Latin in its strength, and the Californian owes it to the leaven of Latin blood.

THE 'FRISCO RESTAURANTS.

With such a people life was always gay. If they did not show it on the streets, as do the people of Paris, it was because the winds made open cafes disagreeable at all seasons of the year. The gayety went on indoors or out on the hundreds of estates that fringed the city. It was noted for its restaurants. Perhaps people who cared not how they spent their money could get the best they wished, but for a dollar down to as low as fifteen cents the restaurants furnished the best fare to be had anywhere at the price.

The country all about produced everything that a cook needs, and that in abundance—the bay was an almost untapped fish-pond, the fruit farms came up to the very edge of the town, and the surrounding country produced in abundance fine meats, all cereals and all vegetables.

But the chefs who came from France in the early days and liked this land of plenty were the head and front of it. They passed their art to other Frenchmen or to the clever Chinese. Most of the French chefs at the biggest restaurants were born in Canton, China. Later the Italians, learning of this country where good food is appreciated, came and brought their own style. Householders always dined out one or two nights of the week, and boarding houses were scarce, for the unattached preferred the restaurants. The eating was usually better than the surroundings.

THE FAMOUS POODLE DOG.

Meals that were marvels were served in tumbledown little hotels. Most famous of all the restaurants was the Poodle Dog. There have been no less than four restaurants of this name, beginning with a frame shanty where, in the early days, a prince of French cooks used to exchange recipes for gold dust. Each succeeding restaurant of the name has moved farther downtown; and the recent Poodle Dog stands—or stood—on the edge of the Tenderloin in a modern five-story building. And it typified a certain spirit that there was in San Francisco.

On the ground floor was a public restaurant where there was served the best dollar dinner on earth. It ranked with the best and the others were in San Francisco. Here, especially on Sunday night, almost everybody went to vary the monotony of home cooking. Every one who was any one in the town could be seen there off and on. It was perfectly respectable. A man might take his wife and daughter there.

On the second floor there were private dining rooms, and to dine there, with one or more of the opposite sex, was risque but not especially terrible. But the third floor—and the fourth floor—and the fifth! The elevator man of the Poodle Dog, who had held the job for many years and never spoke unless spoken to, wore diamonds and was a heavy investor in real estate.

There were others as famous in their way—Zinkaud's, where, at one time, every one went after the theatre, and Tate's, which has lately bitten into that trade; the Palace Grill, much like the grills of Eastern hotels, except for the price; Delmonico's, which ran the Poodle Dog neck and neck in its own line, and many others, humbler, but great at the price.

THE BOHEMIAN CLUB.

To the visitor who came to see the city and who put himself in the hands of one of its well-to-do citizens for the purpose, the few days that followed were apt to be a whirl of mirth and sight-seeing, made up of breakfasts, luncheons, dinners, drives, little trips across the bay, dashes down the peninsula to the polo and country clubs, hours spent in Bohemia, trips around the world among all the races of the habitable globe, all of whom had their colonies in this most cosmopolitan of American cities.

In club life the Bohemian stood first and foremost, the famous club whose meeting place, with all its art treasures, is now a heap of ashes, but which was formerly 'Frisco's head-centre of mirth. Founded by Henry George, the world-famous single tax advocate, when he was an impecunious scribbler on the San Francisco *Post,* it grew to be the choicest place of resort in the Pacific metropolis.

Within its walls the possession of dollars was a bar rather than an "open sesame," the master key to its circles being the knack of telling a good story or the possession of quick and telling wit. Fun-making was the rule there, and the only way to escape being made its victim was the power to deliver a ready and witty retort. In this home of good fellowship all the artists, actors, wits, literati, fiddlers, pianists and bon vivants were members.   Here an impoverished painter could square his grill and buffet account by giving the club a daub to hang on its walls.   Here in days of old the Sheriff used to camp regularly once a month until the members rustled up the money to replevin the furniture.   But these days of poverty passed away, and in later years the club came to know prosperity beyond the dreams of the good fellows who founded it.

THE WICKEDEST AND GAYEST.

The Bohemian is gone, but the spirit that founded and made it still exists, and we may look to see it rise, like the phœnix, from its ashes.

San Francisco was often called the wickedest city in America. It was hardly that, it was simply the gayest.   It was not the home of purity; neither is any other city.   What other cities do behind closed doors San Francisco did not hesitate to do in the open.

In Eastern cities the police have driven vice into tenements, lodging houses and apartments.   San Francisco did not do that. She had certain quarters where, according to unwritten law, vice was allowed to abide, and she did not try to hide the fact that it could be found there. She was not secretly immoral; she was frankly unmoral.

She did not believe in driving her vice from the open where it could be recognized and controlled—prevented from doing any more harm than it was possible to stop—into districts of the city where good people dwell and purity would feel its contaminating influence. There were regions in which the respectable never set foot, haunts of acknowledged vice which for virtue to enter would be to lose caste.

As for its gayety, San Francisco was proud of the reputation of being the Paris of America. Its women were beautiful, and they knew it. They liked to adorn their beauty with fine clothes and peacock along the streets on matinee days. If you asked a San Francisco girl why she wore such expensive clothes, she would say, frankly, "Because I like to have the men admire me," and she would see no harm in saying it. There was very little sham about the San Francisco women. Their men understood them and worshiped them. They bore themselves with the freedom that was theirs by right of their heritage of open-air living, the Bohemian atmosphere they breathed, the unconventional character of their surroundings. Their figures were strong and well moulded, their faces bloomed with health like the roses in their gardens. They drew the wine of laughter from their balmy California air. Sorrow and trouble sat lightly on their shoulders.

There was no end of enjoyments. After the theatre they would go to Zinkand's, Tate's, the Palace or some other of the many places of resort, for a snack to eat and a spell under the music, which was to be heard everywhere.

Another part of the gay life of the city was for a private dance to keep going all night in a fashionable residence, and at daylight, instead of everybody going to bed, to jump into automobiles or carriages or take the trolley cars and whizz off to the beach for a dip in the cold salt water pool at Sutro's baths, and then, with ravenous appetites, sit down on the Cliff House balcony to an open-air breakfast while watching the ships sail in and out at the Golden Gate and hearing the seals barking on the rocks. After that home and to rest.

AN ALL-NIGHT TOWN.

The city never went to sleep altogether. It was "an all-night" town. Few of the restaurants ever closed, none of the saloons did. Always during the whole twenty-four hours of the day there was "something doing" in the Tenderloin. No hour of the night was ever free of revelry. It was marvelous how they kept it up. The average San Franciscon could stay awake all night at a card game, take a cold wash and a good breakfast in the morning, and go straight downtown to business and feel none the worse for it.

It was a gay town, a captivating, piquant, audacious, but not especially wicked city. A Frenchy, a risque city it might justly have been called, but it was not wicked in the sense that sordid vice, vulgar crime and wretched squalor constitute wickedness.

It was a lovable place that everybody longed to get back to, once having been there. A woman, leaving it for years, watched it from the ferryboat, and, weeping, said, "San Francisco, oh, my San Francisco, I am leaving thee."

Will those who left it after the fire ever get back to their old city again? We have already expressed our doubt of this. The old San Francisco is probably gone, never to return. The new San Francisco will be a cleaner, saner and safer city, destitute of its rookeries, its tenements and its Chinatown. It will be a greater and more sightly city than that of the past, but to those who knew and loved the old San Francisco—San Francisco the captivating, the maddest, gayest, liveliest and most rollicking in the country—there must be something impressibly sad to its old inhabitants in the reflection that the new city of the Golden Gate can never be quite the same as the haven of their early affections.

CHAPTER XIII.

Plans to Rebuild San Francisco.

ALMOST as soon as the terrible conflagration had been checked and gotten under control by the heroic efforts of the soldiers and firemen, a little group of the leading citizens of the desolated city had met in the office of Mayor Eugene E. Schmitz and had begun to plan the restoration of their municipality. It was an admirable courage, bred in the stock of those men who in 1849 left comfortable homes in the East to seek their fortune in the Golden State, that inspired the loyal leaders of the present day citizens to provide with far-seeing eyes for the rebuilding of their homes and business houses with more orderly precision after the fire than had been possible during the hustle of early days in a new city.

The old San Francisco was no more, and never could be recalled save as a memory. The local color, atmosphere, that which might be termed the feeling of the old city, vanished with the clustered houses, as rich in tradition as the ancient missions in whose cloisters worshiped the Spanish padre "before the Gringo came." Heart-rending as it was to the citizens who loved their homes and haunts to see them disappear into smoke, there was an attraction about the city of the Golden Gate which endeared it to all Americans.

One of San Francisco's charms was in its defiance of precedent. There were hills to be conquered, and San Francisco's expanding traffic hurled itself at the face of them. It went up and up, with no thought of finding a way around. So it happened that on some of the streets the steepness was too great for horses. In the centre there are cable roads, and on either side of the rails grass grows through the cobbles. The earlier structures on the level were put together in haste. For the most part they remained essentially unchanged until they fell with a crash. True, they had become stained by time, unkempt, dwarfed by new neighbors, but nobody desired to efface them. Away from the business section houses appeared on the various hills, perched precariously near the brink; houses reached by long flights and grown over with roses. The bathing fogs touched them with gray. Moss grew on their roofs. In the little, lofty yards calla lilies bloomed with the profusion of weeds. The natural beauty of the site, the quaintness of the commercial and social development of which it became the centre, attracted the poet and the artist. It incited them to paint the attractions and to sing the praises of their chosen home.

But the loyal sons of those brave pioneers who founded the metropolis were not in the least daunted by the problem of raising from its ruins the whole vast number of dwellings and business houses. The leaders of the people, the men who had been identified with San Francisco since its early days, and whose great fortunes were almost swept away by the cataclysm, lent courage to all the wearied thousands by firm statements of their optimism.

James D. Phelan, former Mayor of the city and one of its richest capitalists, immediately announced his intention of rebuilding his properties at Market and O'Farrell Streets, in the heart of the ruined business district. William H. Crocker, one of the heaviest losers, a nephew of Charles Crocker, who founded the Central Pacific Railroad with Collis P. Huntington, Leland Stanford and others,

SAN FRANCISCO'S CITY HALL.

This photograph shows what was left of the recently completed $7,000,000 City Hall after earthquake and fire had demolished it.

THE BURNING OF NOB HILL.

View looking toward the blazing business district. Some of San Francisco's most magnificent palaces are visible on the right, while the homes of the poor can be seen in the left foreground.

ON THE BEACH SOUTH OF THE CLIFF HOUSE.
Showing the hotel and the Sutro Baths. Many refugees escaped to the beach west of San Francisco and remained idly waiting for the conflagration to burn itself out. Notwithstanding the calamity, they eagerly read the special bulletins issued by the newspapers.

ONE OF THE RUINED PALACES ON NOB HILL.

This photograph shows all that was left of the magnificent home of Charles Crocker, one of San Francisco's millionaires, after earthquake and fire had done their deadly work.

stated emphatically that he would put his shoulder to the wheel. On receiving the first news of the disaster, and before he knew what his losses would amount to, he said:

"Mark my words, San Francisco will arise from these ashes a greater and more beautiful city than ever. I don't take any stock in the belief of some people that investors and residents will be panicky and afraid to build up again. This calamity, terrible as it is, will mean nothing less than a new and grander San Francisco. It is preposterous to suggest the abandonment of the city. It is the natural metropolis of the Pacific coast. God made it so. D. O. Mills, the Spreckels family, everybody I know, have determined to rebuild and to invest more than ever before. Burnham, the great Chicago architect, has been at work for a year or more on plans to beautify San Francisco. Terrible as this destruction has been, it serves to clear the way for the carrying out of these plans. Why, even now we are figuring on rebuilding. More than that, I am confident that, except for what fire has absolutely laid waste, it will be found that the buildings are less injured than was supposed. Plastering, ornamental work, glass and more or less loose material has been shaken down, but the framework, I am sure, will be found intact in many big buildings."

D. Ogden Mills, of New York, who owned enormous properties in the stricken city, was equally confident.

"We will go ahead," said he, "and build the city, and build it so that earthquakes will not shake it down and so fire will not destroy it, and we will have a water system which will enable us to draw water from the sea for fire extinguishing service and other municipal purposes. We will thus have less to fear from the destruction of the land mains. The whole point with all of us who own property down there is that we have to build. To let it lie idle, piled with its ruins, would mean the throwing away of money, and I am sure none of us intends to do that. The city will go up like Baltimore did, and Galveston, and Charleston, and Chicago, and there will be no lack of capital. California spirit and California enterprise, which are always associated with the State of California, will rise superior to this calamity."

George Crocker, elder brother of William H. Crocker; Archer M. Huntington, son of Collis P. Huntington; Mrs. Herman Oelrichs, Mrs. W. K. Vanderbilt, Jr., members of the wealthy Spreckels family and others all expressed, before the great conflagration had ceased burning, the confident expectation that the city would rise, Phœnix-like, from its ashes and become more beautiful and prosperous than it had ever been in the past.

So complete was the calamity that the Government of the United States lent a hand in the earliest work of restoration. On April 20th, two days after the earthquake, Congress took immediate steps to repair or replace all the public buildings damaged or destroyed in San Francisco. The willingness of Congress to assist those in need of work by immediately beginning the reconstruction of the Federal buildings was indicated when Senator Scott, chairman of the Committee on Public Buildings and Grounds, introduced a resolution calling upon the Secretary of the Treasury for full information as to the exact condition of the various government buildings in San Francisco, and instructing him to submit an estimate showing the aggregate sum needed to repair or rebuild them. The resolution suggested that steel frames be used in any new buildings. This resolution was adopted.. It was soon learned that the new Post Office, the Mint and the old Customs House were practically undamaged. The branch of the United States Mint, on Fifth Street, and the new Post Office at Seventh and Mission Streets, were striking examples of the superiority of workmanship put into Federal buildings. The old Mint building, surrounded by a wide space of pavement, was absolutely unharmed. The Mint made preparations to resume business at once. The Post Office building also was virtually undamaged by fire. The earthquake shock did some damage to the different entrances to the building, but the walls were left standing in good condition. President Roosevelt also sent a message to Congress asking that $300,000 be at once appropriated to finish the Mare Island Navy Yard, in order that employment might be given to the many workmen who were in extreme need of money for the necessities of life.

It was a most fortunate circumstance that the property records in the Hall of Records were unharmed either by earthquake or fire. Endless disputes and litigation over the questions of ownerships would undoubtedly have otherwise impeded the work of those sincerely anxious to repair their shattered fortunes and opened the way for the unscrupulous to take unfair advantage of the general chaos.

But the temper of the people was such that only the boldest would have dared to use trickery for his own ends. Every man stood at the side of his neighbor working for himself and for the good of all. Before the embers were cool the owners of some of the damaged skyscrapers gave commands to proceed instantly with their reconstruction. The Spreckels Building, the Hayward Building, the St. Francis Hotel, the Merchants' Exchange and structures that permitted it were ordered rushed into shape as quickly as possible. And already contracts had been drawn up for other steel-frame buildings to be erected with all speed. Many substantial business men and property owners of San Francisco were in consultation with the architects within a few days. While the work of clearing away the debris went forward, a corps of draughtsmen was busily occupied preparing plans for the new buildings to adorn the city.

Mayor Schmitz telegraphed to the Mayors of all leading cities, inquiring how many architects or architectural draughtsmen could be induced to leave for San Francisco at once., and hundreds of young men immediately responded to the call. Experts of the several great contracting companies hurried to the scene and were ready to deposit material and labor on the ground for the work of restoration. Daniel H. Burnham, a leading architect of Chicago, who had previously drawn plans for beautifying the city, was summoned to superintend the work.

All the horses, mules and wagons obtainable were immediately pressed into service to remove the debris and clear the streets so that traffic could be resumed. Within a week after the first earthquake shock trolley cars were running in the principal streets, telephone communication had been re-established in the most needed quarters, electric lights were available and business had begun again on a limited scale.

Yet, in spite of the indomitable courage of the citizens and the efficient labor of the public officers and the utility companies, an enormous amount of work remained. Virtually every bank in San Francisco had to be rebuilt. Only the Market Street National Bank was left nearly undamaged. An official list of the condition of the school buildings throughout the city showed that twenty-nine school buildings were destroyed and that forty-four were partially, at least, spared. Many of the latter were so damaged that they had to be either pulled down or thoroughly repaired, and arrangements were made to resume the short term in tents erected in the parks, where thousands of the homeless had already found temporary shelter. With these two vital classes of public institutions prepared to care for the demands about to be made on them, confidence was not lacking in other parts. Most of the foundries and factories near the water front and south of Market Street immediately called in all their employees and began to clear away the wreckage and make ready for continuing business. Great credit is due to the news-papers, nearly all of which continued their daily issues without interruption, although their buildings, with offices and printing plants, were entirely destroyed by the flames which followed the earthquake. Those whose premises were early threatened with destruction betook themselves to Oakland, seven miles distant across the bay, and published their sheets from the establishments of the Oakland papers. A thorough inspection shows that comparatively little damage was done in the vicinity of the Cliff. The Cliff House, which was at first reported to have been hurled into the sea, not only stood, but the damage sustained by it from the earthquake was slight. The famous Sutro baths, located near the Cliff House, with the hundreds of thousands of square feet of glass roofing, also were practically unharmed. Only a few of the windows in the Sutro baths and the Cliff House were broken, and the lofty chimney of the pumping plant of the former establishment was cracked only a trifle. When the situation was finally summed up, however, nearly three-fourths of the city had to be rebuilt or remodeled, and the cost of doing this was enough to appal the strongest hearts.

Financially the prospect was encouraging. Not a bank lost

the contents of its fireproof vaults and remained practically un-harmed, so far as credit was concerned.

For a number of days it was impossible to open any strong boxes on account of the great heat which the thick walls retained, and this naturally caused some embarrassment and lack of ready money. Nearly all of them, however, had strong connections in Eastern cities and large balances to their credit in other banks of America and Europe. They were also favored by the fact that the United States Mint and the Sub-Treasury held between them some $245,000,000 in ready money. The Secretary of the Treasury im-mediately deposited $10,000,000 to the credit of the local banks, and financiers of the great business centres of the country added to public confidence by prompt statements that they would facilitate the reconstruction of the city by a liberal advancement of funds.

One prominent Eastern capitalist expressed the general con-viction in the following words:

"No great city, unless it dried up entirely from lack of commer-cial life blood, was ever annihilated by such a disaster as that of San Francisco. Pompeii and Herculaneum were not great cities in the first place, and in the second, they were completely covered, smothered as it were, with the ashes and molten lava of the adjoining volcano, and nearly all of their inhabitants perished. If it be ad-mitted that three-fourths of the superstructures, so to speak, of San Francisco, estimated according to valuation, is destroyed, we have yet the fact remaining that the lives of only about one four-hundredth of its population have been lost.

"San Francisco was not merely land and the buildings erected upon it, but it was people, and one of the most active, most hopeful, most vivacious human communities on the face of the earth. You cannot long discourage such a community, unless you wipe out three-fourths of its members. Will San Francisco rise again? Most certainly it will. Galveston and Baltimore, not to mention Charleston, Boston and Chicago, showed the spirit of material resurrection in American communities, sore-smitten by calamity. After Galveston had been made a desert of sand and debris, there were predictions that it would never rise again. What was the outcome? A finer Galveston than before, and finer than many years of slow improvement in the natural course would have made it. Baltimore is busier commercially than it was before the great fire.

"San Francisco is exceedingly fortunate in the fact that its moneyed institutions remain strong, with abundant supplies of funds. It is true that many of them undoubtedly hold large numbers of real estate mortgages as securities for loans, and that much of the property thus represented is now in ashes. But with care and an accommodating spirit practically all of those mortgaged can be so nursed that they will be made absolutely good. The banks will be found to be only too eager to afford new loans which will enable realty owners to rebuild. You will see San Francisco rise a more splendid city than ever, and better prepared to resist future earthquake shocks. Because it has had this dreadful visitation is no reason for apprehension that another like it will come within the life of the present generation, or two or three after. The destruction of Lisbon in the middle of the eighteenth century and its subsequent immunity from seismic damage is a reassuring example."

The municipality was in excellent financial condition to meet and rise above the extraordinary needs of the situation. It had a bonded debt of only $4,245,100, while its realty valuation was $402,127,261 and its personalty $122,258,406. The question of issuing further amounts of bonds was therefore one of the first measures considered by Mayor Schmitz and his co-workers, and an appeal was made to the Federal Government to guarantee the proposed loans, so that the most urgent work which lay in the city's province could be undertaken at once and without an excessive burden of interest.

The vast insurance loss was divided among 107 companies, and, though only a little more than half the damage was covered by policies, the total swelled toward the colossal sum of $150,000,000. Several of the largest companies were seriously crippled by the disaster and some were forced into liquidation. To the great relief of the entire country, nevertheless, the financial situation was not severely affected, and there was every reason to believe that the great bulk of the insurance would be paid.

CHAPTER XIV.

The Earthquake Wave Felt Round the Earth.

THE outbreak of earth forces at San Francisco did not stand alone. There were others elsewhere at nearly the same time, the whole seeming to indicate a general disturbance in the interior of the earth's crust. Some scientists, indeed, declared that no possible connection could exist between the eruption of Mount Vesuvius and the earthquake at San Francisco, but others were inclined to view certain facts in regard to recent seismic and volcanic activity as, to say the least, suggestive.

As to the actual cause of the California earthquake, the wisest confession we can make is that of ignorance, there being almost as little known as to the origin, period and coming of earthquakes as when Pliny wrote 1,800 years ago. The Roman observer knew that the tremor passed like a wave through the surface of the earth; he knew that it had a given direction, and he knew that certain regions were rife with seismic disturbance. More he could not say, and when this is said all has been said that is known to-day.

Setting aside these general considerations, let us return to the question of the disaster at San Francisco on that fatal morning of April 18th. The shock did not come unexpectedly. A month previous there had been a severe earthquake in the Island of Formosa, and many lives were lost there, while an enormous amount of damage was done. Only a few days before the event in San Francisco there was another earthquake in the same island. Still greater havoc was caused by it than by the earthquake in March, but fewer lives were lost, the reason being that the people were warned in time.   Early in April the eruption of Mount Vesuvius reached its height and devastated the country around the volcano, covering an enormous territory with ashes, and caused the loss of hundreds of lives.

On Tuesday night, April 17th, word was received from Piatigorsk, Circassia, that there had been two severe earthquake shocks the previous day in Northern Caucasia.   The same night a telegram from Madrid said that the newspapers there reported that the long-dormant volcano on Palma, the largest of the Canary Islands, was showing signs of eruption, columns of smoke issuing from the crater.

WIDESPREAD EARTH TREMORS.

While scientists as a rule doubt that there was any connection between these volcanic phenomena and the earthquake at San Francisco, yet reports from the Mount Weather observation station in Virginia, a few miles from Washington, show that the eruptions of Vesuvius acted on the magnetic instruments by electro-magnetic waves in such a way as to disturb the electrical potentials at that place.   Be this as it may, there is one remarkable circumstance in regard to all this activity.   All the places mentioned—Formosa, Southern Italy, Caucasia, and the Canary Islands—lie within a belt bounded by lines a little north of the fortieth parallel and a little south of the thirtieth parallel.   San Francisco is just south of the fortieth parallel, while Naples is just north of it.   The latitude of Calabria, where the terrible earthquakes occurred in 1905, is the same as that of the territory affected by the recent earthquake in the United States. This may or may not have some bearing on the question.

Whatever be thought of all this, one thing is certain, the earthquake which laid San Francisco in ruins was felt the world over, wherever there were instruments in position to detect and record it. The seismograph in the government observatory at Washington showed that the first wave, on April 18th, came at 8.19—equivalent to 5.19 at San Francisco; that at 8.25 there was a stronger wave motion, and that from 8.32 to 8.35 the recording pen was carried off the paper. The vibrations did not entirely cease until 12.35 P. M., during this period there having been nearly half an inch of to and fro motion in the surface of the earth.

RECORDS OF FOREIGN OBSERVATIONS.

From far away New Zealand, on the same date, the government seismograph at the capital, Wellington, recorded seismic waves that apparently passed round the earth five times at intervals of about four hours each.

Across the Atlantic, at Heidelberg, in Germany, the records showed vibrations lasting one hour. At Sarayevo, in Bosnia, there was a sharp shock at 11 A. M., undulating from west to east. At Funfkirchen, in Hungary, at Laibach, in Austria, in the Isle of Wight, off the coast of England, and all through Italy, from north to south, the shocks were felt.

At Hancock, Mich., a shock was felt on April 19th a mile below the surface in the Quincy mine of such severity that one man was killed and four injured by a fall of rock loosened by the trembling of the earth.   There is no evidence, however, that this had any connection with the California disaster, the dates not coinciding.

Turning to the Far East, across the Pacific, seismographs in the Imperial University of Tokio showed that the earthquake was felt there eleven minutes later than in San Francisco, and similar instruments in Manila detected the arrival of the seismic waves twenty minutes after the San Francisco shock.   In this there was a slight difference in time compared with Tokio, but, considering the distance, near enough to prove that the disturbances came from the same source.

Not until the day following was any noticeable disturbance felt in Honolulu, but on April 19th shocks were plainly felt for six minutes and the water in the harbor rose rapidly.   Panic seemed imminent just before the shocks subsided.   While earthquakes are by no means infrequent in these islands, this was more severe than any recorded in recent years, causing buildings to sway to and fro and partly demolishing some of frail construction.

If, as the majority of men qualified to discuss earthquakes seem to think, the San Francisco earthquake had no connection with volcanic action, but was caused by what is technically known as a "fault" in the formation of the crust of the earth, it seems easy enough to account for these wave motions travelling round the earth.   How widely this may really have made itself felt it is not possible to say.   Several of the great earthquakes in Japan have been recorded in the seismographs of the observatories on every continent and in Australia, showing that in severe disturbances of this kind the whole surface strata quiver, alike under the oceans and over the continents and islands.   At the time of a shock, of course, half of the world is in darkness and asleep. This is taken to account for the fact that so far only a few observatories have reported catching the San Francisco vibrations.

The instruments invented for the recording of the motions of the earth's crust are looked upon by scientists as the most delicate of all machines. So highly sensitive are they, indeed, that the very slightest vibratory motion is recorded perfectly. Even the tread of feet cannot escape this instrument if sufficient to cause a vibration.

There are three classes of instruments for the automatic recording of earth tremors, each with its own particular function. First is the seismoscope, which will merely detect and record the fact that there has been such a tremor. Some of these are so equipped as to indicate the time of the disturbance.

Second, is the seismometer, the function of which is to measure the maximum force of the shock, either with or without an indication of its direction. The third instrument is the seismograph, which is so arranged that it will accurately record the number, succession, direction, amplitude and period of successive oscillations. This last instrument is by far the most delicate of the three.

In the construction of this earthquake recording machine the maker must so suspend a heavy body that when its normal position is disturbed in the most infinitesimal degree no reactionary force will be developed tending to restore it to its original position. The inventor has never been found who could accomplish this suspension of a body to perfection. The seismograph of to-day, however, has reached a stage of perfection where close approximations are obtained in the records made.

CAMPING ON THE OUTSKIRTS OF SAN FRANCISCO.

Rich and poor alike made temporary shelters for themselves and furnished them with such things as they had saved from the fire until more comfortable quarters could be provided to take the place of their ruined homes.

RUINS OF THE VALENCIA HOTEL, SAN FRANCISCO.
Forty people were killed by the collapsing of this structure at the first earthquake shock. It was a four-story building. Three stories settled into the earth.

Copyright, 1906. Judge Co.
THE BLAZING BUSINESS DISTRICT OF SAN FRANCISCO.
An awe-inspiring sight of millions of dollars and the labor of half a century being destroyed
by the fiery element which was shaken into life all over the city by the
earthquake shock of April 18th.

STANDING IN LINE FOR RATIONS.

This photograph, taken in Golden Gate Park, shows rich and poor alike patiently waiting their turn to secure some of the coveted packages of food which were sent to relieve their sufferings from all parts of the country.

REFUGEES LEAVING THE FERRY HOUSE.
Twisted and toppling in the consuming heat, the tall tower of the Ferry House stood like a
silhouette against the lurid background of flame.

MILLIONAIRES ESCAPING IN AUTOMOBILES.
The rich piled their valuables in automobiles, wagons, or any conveyance they could find, and fled to the parks and open spaces in the western part of the city.

BERKLEY BLDG.   OAKLAND BLDG.   POSTAL TEL.   CHRONICLE   PALACE   EXAMINER   CALL BLDG.   GRAND OPERA HOUSE.
BLDG.   BLDG.   HOTEL   BLDG.

Copyright, 1906, by W. E. Scull.
THE BURNING HEART OF SAN FRANCISCO.
A heart-rending panorama of the great city by the Golden Gate, made at the height of the
conflagration.

DYNAMITING A MILE OF PALACES AT VAN NESS AVENUE.
Soldiers and firemen blew up sixteen blocks of magnificent residences and partially succeeded
in staying the devouring progress of the flames by destroying their fuel.

THE MISSION OF SAN JUAN CAPISTRANO.

These picturesque cloisters have long stood the ravages of time and many previous slighter earthquake shocks.

SAN FRANCISCO BEFORE THE FIRE.

Looking southeast from an eminence in Union Square, a characteristic view of the business section of the city. The shipping and the bay appear in the background.

Copyright, 1906, by W. E. Scull.

THE EVACUATION OF CHINATOWN.

The thousands of Chinese who inhabited their own quarter in the city fled to the parks and the ferries with all their movable possessions.

CAMPING IN THE GOLDEN GATE PARK.

Thousands of homeless families set up cooking stoves in the park and made arrangements to spend days and weeks under the roof of the sky.

AUTOMOBILES AND CARRIAGES OVERTAKEN BY FIRE.

In the foreground is seen the skeleton of an automobile of some wealthy San Franciscan who was forced to abandon his machine in the mad panic of flight from the burning city. The remains of other carriages appear on the opposite side of the street.

THE HAVOC OF THE EARTHQUAKE.
St. Dominick's Church, on Stevens, near Pine Street, was destroyed by earthquake
in a district not reached by the fire. One of the cupolas which adorned
the building was entirely destroyed, while the other remained
suspended like a bell on the top of the spire.

A BANK IN OAKLAND WRECKED BY EARTHQUAKE.

The heavy cornice and sheathing of this building crashed down at the first shock of April 18, 1906, and killed four men.

HUGE SKY-SCRAPERS THREATENED BY THE SWIFTLY ADVANCING FLAMES
IN SAN FRANCISCO ON APRIL 19TH.

The Hayward Building at the left, Merchants' Exchange in the central background and the
Mills Building at the right. This remarkable photograph shows the
Demon of fire at the height of its destructive violence.

Copyright, 1906, National Press Association.
THE DESOLATED HEART OF SAN FRANCISCO.
A photograph in the business section of the city after earthquake and fire had done their
deadly work. Gaunt walls and heaps of debris show how complete was the destruction.

VIEW FROM A COUNTRY PLACE IN SANTA CLARA COUNTY, CALIFORNIA.
It is almost possible to live in the open air the whole year round. Many magnificent country places are scattered through the coast region.

A STREET IN CHINATOWN.
The largest Chinese colony in America occupied the northeastern section
of San Francisco. Its flimsy buildings made it an
easy prey to the flames.

TWO OF THE RUINED RESIDENCES.
Below, the magnificent home of Charles Crocker, one of San Francisco's
wealthiest men. Above, the Stanford house.

TWO PANORAMAS OF SAN FRANCISCO.
Below, taken from a western hill, showing the Hebrew Synagogue in the
foreground. Above, a bird's-eye view of the
business portion of the city.

LELAND STANFORD, Jr., UNIVERSITY.
The famous institution at Palo Alto, thirty-four miles south of San Francisco, endowed with $30,000,000 by the late Senator Stanford in memory of his son. All the costly buildings were demolished by the earthquake except one.

SEAL ROCKS AND THE CLIFF HOUSE.

This famous hostelry was in imminent danger of destruction by the earthquake.

PLACES OF REFUGE FOR THE HOMELESS.
One of the beautiful homes in the western part of San Francisco and a
scene in Golden Gate Park, where thousands of fugi-
tives slept in tents and rude shelters.

THE GOLDEN GATE FROM THE CLIFF HOUSE.
The famous entrance to San Francisco Bay. The outlet of a vast trade to the Orient and the Philippines.

THE PRECIPITOUS ROCKS WHERE THE CLIFF HOUSE STANDS.
A view from the site of this famous hotel, looking south along the beautiful shore of the Pacific.

A CATALOG OF SELECTED
DOVER BOOKS
IN ALL FIELDS OF INTEREST

A CATALOG OF SELECTED DOVER
BOOKS IN ALL FIELDS OF INTEREST

100 BEST-LOVED POEMS, Edited by Philip Smith. "The Passionate Shepherd to His Love," "Shall I compare thee to a summer's day?" "Death, be not proud," "The Raven," "The Road Not Taken," plus works by Blake, Wordsworth, Byron, Shelley, Keats, many others. Includes 13 selections from the Common Core State Standards Initiative. 112pp. 0-486-28553-7

ABC BOOK OF EARLY AMERICANA, Eric Sloane. Artist and historian Eric Sloane presents a wondrous A-to-Z collection of American innovations, including hex signs, ear trumpets, popcorn, and rocking chairs. Illustrated, hand-lettered pages feature brief captions explaining objects' origins and uses. 64pp. 0-486-49808-5

ADVENTURES OF HUCKLEBERRY FINN, Mark Twain. Join Huck and Jim as their boyhood adventures along the Mississippi River lead them into a world of excitement, danger, and self-discovery. Humorous narrative, lyrical descriptions of the Mississippi valley, and memorable characters. 224pp. 0-486-28061-6

ALICE STARMORE'S BOOK OF FAIR ISLE KNITTING, Alice Starmore. A noted designer from the region of Scotland's Fair Isle explores the history and techniques of this distinctive, stranded-color knitting style and provides copious illustrated instructions for 14 original knitwear designs. 208pp. 0-486-47218-3

ALICE'S ADVENTURES IN WONDERLAND, Lewis Carroll. Beloved classic about a little girl lost in a topsy-turvy land and her encounters with the White Rabbit, March Hare, Mad Hatter, Cheshire Cat, and other delightfully improbable characters. 42 illustrations by Sir John Tenniel. A selection of the Common Core State Standards Initiative. 96pp. 0-486-27543-4

THE ARTHUR RACKHAM TREASURY: 86 Full-Color Illustrations, Arthur Rackham. Selected and Edited by Jeff A. Menges. A stunning treasury of 86 full-page plates span the famed English artist's career, from *Rip Van Winkle* (1905) to masterworks such as *Undine, A Midsummer Night's Dream,* and *Wind in the Willows* (1939). 96pp. 0-486-44685-9

THE AWAKENING, Kate Chopin. First published in 1899, this controversial novel of a New Orleans wife's search for love outside a stifling marriage shocked readers. Today, it remains a first-rate narrative with superb characterization. New introductory note. 128pp. 0-486-27786-0

BASEBALL IS . . .: Defining the National Pastime, Edited by Paul Dickson. Wisecracking, philosophical, nostalgic, and entertaining, these hundreds of quips and observations by players, their wives, managers, authors, and others cover every aspect of our national pastime. It's a great any-occasion gift for fans! 256pp. 0-486-48209-X

THE CALL OF THE WILD, Jack London. A classic novel of adventure, drawn from London's own experiences as a Klondike adventurer, relating the story of a heroic dog caught in the brutal life of the Alaska Gold Rush. Note. 64pp. 0-486-26472-6

CANDIDE, Voltaire. Edited by Francois-Marie Arouet. One of the world's great satires since its first publication in 1759. Witty, caustic skewering of romance, science, philosophy, religion, government — nearly all human ideals and institutions. A selection of the Common Core State Standards Initiative. 112pp. 0-486-26689-3

THE CARTOON HISTORY OF TIME, Kate Charlesworth and John Gribbin. Cartoon characters explain cosmology, quantum physics, and other concepts covered by Stephen Hawking's *A Brief History of Time.* Humorous graphic novel–style treatment, perfect for young readers and curious folk of all ages. 64pp. 0-486-49097-1

THE CHERRY ORCHARD, Anton Chekhov. Classic of world drama concerns passing of semifeudal order in turn-of-the-century Russia, symbolized in the sale of the cherry orchard owned by Madame Ranevskaya. Showcases Chekhov's rich sensitivities as an observer of human nature. 64pp. 0-486-26682-6

A CHRISTMAS CAROL, Charles Dickens. This engrossing tale relates Ebenezer Scrooge's ghostly journeys through Christmases past, present, and future and his ultimate transformation from a harsh and grasping old miser to a charitable and compassionate human being. 80pp. 0-486-26865-9

CRIME AND PUNISHMENT, Fyodor Dostoyevsky. Translated by Constance Garnett. Supreme masterpiece tells the story of Raskolnikov, a student tormented by his own thoughts after he murders an old woman. Overwhelmed by guilt and terror, he confesses and goes to prison. A selection of the Common Core State Standards Initiative. 448pp. 0-486-41587-2

CYRANO DE BERGERAC, Edmond Rostand. A quarrelsome, hot-tempered, and unattractive swordsman falls hopelessly in love with a beautiful woman and woos her for a handsome but slow-witted suitor. A witty and eloquent drama. 144pp. 0-486-41119-2

A DOLL'S HOUSE, Henrik Ibsen. Ibsen's best-known play displays his genius for realistic prose drama. An expression of women's rights, the play climaxes when the central character, Nora, rejects a smothering marriage and life in "a doll's house." A selection of the Common Core State Standards Initiative. 80pp. 0-486-27062-9

DOOMED SHIPS: Great Ocean Liner Disasters, William H. Miller, Jr. Nearly 200 photographs, many from private collections, highlight tales of some of the vessels whose pleasure cruises ended in catastrophe: the *Morro Castle, Normandie, Andrea Doria, Europa,* and many others. 128pp. 0-486-45366-9

DUBLINERS, James Joyce. A fine and accessible introduction to the work of one of the 20th century's most influential writers, this collection features 15 tales, including a masterpiece of the short-story genre, "The Dead." 160pp. 0-486-26870-5

THE EARLY SCIENCE FICTION OF PHILIP K. DICK, Philip K. Dick. This anthology presents short stories and novellas that originally appeared in pulp magazines of the early 1950s, including "The Variable Man," "Second Variety," "Beyond the Door," "The Defenders," and more. 272pp. 0-486-49733-X

THE EARLY SHORT STORIES OF F. SCOTT FITZGERALD, F. Scott Fitzgerald. These tales offer insights into many themes, characters, and techniques that emerged in Fitzgerald's later works. Selections include "The Curious Case of Benjamin Button," "Babes in the Woods," and a dozen others. 256pp. 0-486-79465-2

ETHAN FROME, Edith Wharton. Classic story of wasted lives, set against a bleak New England background. Superbly delineated characters in a hauntingly grim tale of thwarted love. Considered by many to be Wharton's masterpiece. 96pp. 0-486-26690-7

FLATLAND: A Romance of Many Dimensions, Edwin A. Abbott. Classic of science (and mathematical) fiction — charmingly illustrated by the author — describes the adventures of A. Square, a resident of Flatland, in Spaceland (three dimensions), Lineland (one dimension), and Pointland (no dimensions). 96pp. 0-486-27263-X

FRANKENSTEIN, Mary Shelley. The story of Victor Frankenstein's monstrous creation and the havoc it caused has enthralled generations of readers and inspired countless writers of horror and suspense. With the author's own 1831 introduction. 176pp. 0-486-28211-2

THE GARGOYLE BOOK: 572 Examples from Gothic Architecture, Lester Burbank Bridaham. Dispelling the conventional wisdom that French Gothic architectural flourishes were born of despair or gloom, Bridaham reveals the whimsical nature of these creations and the ingenious artisans who made them. 572 illustrations. 224pp. 0-486-44754-5

Browse over 10,000 books at www.doverpublications.com

THE GIFT OF THE MAGI AND OTHER SHORT STORIES, O. Henry. Sixteen captivating stories by one of America's most popular storytellers. Included are such classics as "The Gift of the Magi," "The Last Leaf," and "The Ransom of Red Chief." Publisher's Note. A selection of the Common Core State Standards Initiative. 96pp.   0-486-27061-0

THE GOETHE TREASURY: Selected Prose and Poetry, Johann Wolfgang von Goethe. Edited, Selected, and with an Introduction by Thomas Mann. In addition to his lyric poetry, Goethe wrote travel sketches, autobiographical studies, essays, letters, and proverbs in rhyme and prose. This collection presents outstanding examples from each genre. 368pp.   0-486-44780-4

GREAT ILLUSTRATIONS BY N. C. WYETH, N. C. Wyeth. Edited and with an Introduction by Jeff A. Menges. This full-color collection focuses on the artist's early and most popular illustrations, featuring more than 100 images from *The Mysterious Stranger, Robin Hood, Robinson Crusoe, The Boy's King Arthur,* and other classics. 128pp.   0-486-47295-7

HAMLET, William Shakespeare. The quintessential Shakespearean tragedy, whose highly charged confrontations and anguished soliloquies probe depths of human feeling rarely sounded in any art. Reprinted from an authoritative British edition complete with illuminating footnotes. A selection of the Common Core State Standards Initiative. 128pp.   0-486-27278-8

THE HAUNTED HOUSE, Charles Dickens. A Yuletide gathering in an eerie country retreat provides the backdrop for Dickens and his friends — including Elizabeth Gaskell and Wilkie Collins — who take turns spinning supernatural yarns. 144pp.   0-486-46309-5

HEART OF DARKNESS, Joseph Conrad. Dark allegory of a journey up the Congo River and the narrator's encounter with the mysterious Mr. Kurtz. Masterly blend of adventure, character study, psychological penetration. For many, Conrad's finest, most enigmatic story. 80pp.   0-486-26464-5

THE HOUND OF THE BASKERVILLES, Sir Arthur Conan Doyle. A deadly curse in the form of a legendary ferocious beast continues to claim its victims from the Baskerville family until Holmes and Watson intervene. Often called the best detective story ever written. 128pp.   0-486-28214-7

THE HOUSE BEHIND THE CEDARS, Charles W. Chesnutt. Originally published in 1900, this groundbreaking novel by a distinguished African-American author recounts the drama of a brother and sister who "pass for white" during the dangerous days of Reconstruction. 208pp.   0-486-46144-0

HOW TO DRAW NEARLY EVERYTHING, Victor Perard. Beginners of all ages can learn to draw figures, faces, landscapes, trees, flowers, and animals of all kinds. Well-illustrated guide offers suggestions for pencil, pen, and brush techniques plus composition, shading, and perspective. 160pp.   0-486-49848-4

HOW TO MAKE SUPER POP-UPS, Joan Irvine. Illustrated by Linda Hendry. Super pop-ups extend the element of surprise with three-dimensional designs that slide, turn, spring, and snap. More than 30 patterns and 475 illustrations include cards, stage props, and school projects. 96pp.   0-486-46589-6

THE IMITATION OF CHRIST, Thomas à Kempis. Translated by Aloysius Croft and Harold Bolton. This religious classic has brought understanding and comfort to millions for centuries. Written in a candid and conversational style, the topics include liberation from worldly inclinations, preparation and consolations of prayer, and eucharistic communion. 160pp.   0-486-43185-1

THE IMPORTANCE OF BEING EARNEST, Oscar Wilde. Wilde's witty and buoyant comedy of manners, filled with some of literature's most famous epigrams, reprinted from an authoritative British edition. Considered Wilde's most perfect work. A selection of the Common Core State Standards Initiative. 64pp. 0-486-26478-5

JANE EYRE, Charlotte Brontë. Written in 1847, *Jane Eyre* tells the tale of an orphan girl's progress from the custody of cruel relatives to an oppressive boarding school and its culmination in a troubled career as a governess. A selection of the Common Core State Standards Initiative. 448pp. 0-486-42449-9

JUST WHAT THE DOCTOR DISORDERED: Early Writings and Cartoons of Dr. Seuss, Dr. Seuss. Edited and with an Introduction by Rick Marschall. The Doctor's visual hilarity, nonsense language, and offbeat sense of humor illuminate this compilation of items from his early career, created for periodicals such as *Judge, Life, College Humor,* and *Liberty.* 144pp. 0-486-49846-8

KING LEAR, William Shakespeare. Powerful tragedy of an aging king, betrayed by his daughters, robbed of his kingdom, descending into madness. Perhaps the bleakest of Shakespeare's tragic dramas, complete with explanatory footnotes. 144pp. 0-486-28058-6

THE LADY OR THE TIGER?: and Other Logic Puzzles, Raymond M. Smullyan. Created by a renowned puzzle master, these whimsically themed challenges involve paradoxes about probability, time, and change; metapuzzles; and self-referentiality. Nineteen chapters advance in difficulty from relatively simple to highly complex. 1982 edition. 240pp. 0-486-47027-X

LEAVES OF GRASS: The Original 1855 Edition, Walt Whitman. Whitman's immortal collection includes some of the greatest poems of modern times, including his masterpiece, "Song of Myself." Shattering standard conventions, it stands as an unabashed celebration of body and nature. 128pp. 0-486-45676-5

LES MISÉRABLES, Victor Hugo. Translated by Charles E. Wilbour. Abridged by James K. Robinson. A convict's heroic struggle for justice and redemption plays out against a fiery backdrop of the Napoleonic wars. This edition features the excellent original translation and a sensitive abridgment. 304pp. 0-486-45789-3

LIGHT FOR THE ARTIST, Ted Seth Jacobs. Intermediate and advanced art students receive a broad vocabulary of effects with this in-depth study of light. Diagrams and paintings illustrate applications of principles to figure, still life, and landscape paintings. 144pp. 0-486-49304-0

LILITH: A Romance, George MacDonald. In this novel by the father of fantasy literature, a man travels through time to meet Adam and Eve and to explore humanity's fall from grace and ultimate redemption. 240pp. 0-486-46818-6

LINE: An Art Study, Edmund J. Sullivan. Written by a noted artist and teacher, this well-illustrated guide introduces the basics of line drawing. Topics include third and fourth dimensions, formal perspective, shade and shadow, figure drawing, and other essentials. 208pp. 0-486-79484-9

THE LODGER, Marie Belloc Lowndes. Acclaimed by *The New York Times* as "one of the best suspense novels ever written," this novel recounts an English couple's doubts about their boarder, whom they suspect of being a serial killer. 240pp. 0-486-78809-1

MACBETH, William Shakespeare. A Scottish nobleman murders the king in order to succeed to the throne. Tortured by his conscience and fearful of discovery, he becomes tangled in a web of treachery and deceit that ultimately spells his doom. A selection of the Common Core State Standards Initiative. 96pp. 0-486-27802-6

MANHATTAN IN MAPS 1527–2014, Paul E. Cohen and Robert T. Augustyn. This handsome volume features 65 full-color maps charting Manhattan's development from the first Dutch settlement to the present. Each map is placed in context by an accompanying essay. 176pp. 0-486-77991-2

MEDEA, Euripides. One of the most powerful and enduring of Greek tragedies, masterfully portraying the fierce motives driving Medea's pursuit of vengeance for her husband's insult and betrayal. Authoritative Rex Warner translation. 64pp. 0-486-27548-5

THE METAMORPHOSIS AND OTHER STORIES, Franz Kafka. Excellent new English translations of title story (considered by many critics Kafka's most perfect work), plus "The Judgment," "In the Penal Colony," "A Country Doctor," and "A Report to an Academy." A selection of the Common Core State Standards Initiative. 96pp. 0-486-29030-1

METROPOLIS, Thea von Harbou. This Weimar-era novel of a futuristic society, written by the screenwriter for the iconic 1927 film, was hailed by noted science-fiction authority Forrest J. Ackerman as "a work of genius." 224pp. 0-486-79567-5

THE MYSTERIOUS MICKEY FINN, Elliot Paul. A multimillionaire's disappearance incites a maelstrom of kidnapping, murder, and a plot to restore the French monarchy. "One of the funniest books we've read in a long time." — *The New York Times.* 256pp. 0-486-24751-1

NARRATIVE OF THE LIFE OF FREDERICK DOUGLASS, Frederick Douglass. The impassioned abolitionist and eloquent orator provides graphic descriptions of his childhood and horrifying experiences as a slave as well as a harrowing record of his dramatic escape to the North and eventual freedom. A selection of the Common Core State Standards Initiative. 96pp. 0-486-28499-9

OBELISTS FLY HIGH, C. Daly King. Masterpiece of detective fiction portrays murder aboard a 1935 transcontinental flight. Combining an intricate plot and "locked room" scenario, the mystery was praised by *The New York Times* as "a very thrilling story." 288pp. 0-486-25036-9

THE ODYSSEY, Homer. Excellent prose translation of ancient epic recounts adventures of the homeward-bound Odysseus. Fantastic cast of gods, giants, cannibals, sirens, other supernatural creatures — true classic of Western literature. A selection of the Common Core State Standards Initiative. 256pp. 0-486-40654-7

OEDIPUS REX, Sophocles. Landmark of Western drama concerns the catastrophe that ensues when King Oedipus discovers he has inadvertently killed his father and married his mother. Masterly construction, dramatic irony. A selection of the Common Core State Standards Initiative. 64pp. 0-486-26877-2

OTHELLO, William Shakespeare. Towering tragedy tells the story of a Moorish general who earns the enmity of his ensign Iago when he passes him over for a promotion. Masterly portrait of an archvillain. Explanatory footnotes. 112pp. 0-486-29097-2

THE PICTURE OF DORIAN GRAY, Oscar Wilde. Celebrated novel involves a handsome young Londoner who sinks into a life of depravity. His body retains perfect youth and vigor while his recent portrait reflects the ravages of his crime and sensuality. 176pp. 0-486-27807-7

A PLACE CALLED PECULIAR: Stories About Unusual American Place-Names, Frank K. Gallant. From Smut Eye, Alabama, to Tie Siding, Wyoming, this pop-culture history offers a well-written and highly entertaining survey of America's most unusual place-names and their often-humorous origins. 256pp. 0-486-48360-6

PRIDE AND PREJUDICE, Jane Austen. One of the most universally loved and admired English novels, an effervescent tale of rural romance transformed by Jane Austen's art into a witty, shrewdly observed satire of English country life. A selection of the Common Core State Standards Initiative. 272pp. 0-486-28473-5

Browse over 10,000 books at www.doverpublications.com